£1.50

UNITS AND STANDARDS
FOR ELECTROMAGNETISM

D1824333

THE WYKEHAM SCIENCE SERIES

General Editors:

PROFESSOR SIR NEVILL MOTT, F.R.S.
Cavendish Professor of Physics
University of Cambridge

G. R. NOAKES
Formerly Senior Physics Master
Uppingham School

To introduce the present state of science as a university subject to students approaching or starting their university careers is the aim of the Wykeham Science Series.   Each book seeks to reinforce the link between school and university levels, and the main author, a university teacher or research worker distinguished in the field, is assisted by an experienced sixth-form schoolmaster.

UNITS AND STANDARDS
FOR ELECTROMAGNETISM

P. Vigoureux – National Physical Laboratory

WYKEHAM PUBLICATIONS (LONDON) LTD
(A MEMBER OF THE TAYLOR & FRANCIS GROUP)
LONDON AND WINCHESTER
1971

First published 1971 by Wykeham Publications (London) Ltd.

Cover illustration—NPL ' weak field' apparatus for measuring the gyro-magnetic ratio of the proton. (Crown copyright reserved.)

ISBN 0 85109 190 3

Printed in Great Britain by Taylor & Francis Ltd.
10–14 Macklin Street, London, WC2B 5NF

Distribution:

UNITED KINGDOM, EUROPE, MIDDLE EAST AND AFRICA
Chapman & Hall Ltd. (a member of Associated Book Publishers Ltd.), 11 New Fetter Lane, London, E.C.4, and North Way, Andover, Hampshire.

UNITED STATES OF AMERICA, CANADA AND MEXICO
Springer-Verlag New York Inc., 175 Fifth Avenue, New York, New York 10010.

AUSTRALIA AND NEW GUINEA
Hicks Smith & Sons Pty. Ltd., 301 Kent Street, Sydney, N.S.W. 2000.

NEW ZEALAND AND FIJI
Hicks Smith & Sons Ltd., 238 Wakefield Street, Wellington.

ALL OTHER TERRITORIES
Taylor & Francis Ltd., 10–14 Macklin Street, London, WC2B 5NF.

PREFACE

USE of fundamental atomic constants to realize and maintain the units of length and time and of electromagnetism is now well established, and its broad principles, if not its details, should be made known to undergraduate students and to 6th form pupils. The account given here will, it is hoped, be found sufficiently informative by the former, yet readable without too much difficulty by the latter.

The units treated are the metre, second, kilogram, ampere, ohm and volt, and the fundamental constants involved are the speed of light, the gyromagnetic ratio of the proton, and the ratio $2\,e/h$ derived from the Josephson effect in superconductors. As it would be neither practical nor desirable to discuss realization of electric units without explaining also how they are maintained, resistors, Weston cells and associated measuring devices are described in the last two chapters, but only to the extent needed to illustrate their use in equipment discussed earlier.

Suggestions for further reading are offered at the end of most chapters.

Inevitably my long connection with the National Physical Laboratory, UK, has resulted in methods employed and instruments made there being selected for description more often than those used elsewhere. That is not to say that I claim for them superiority over equipment and measurements of other laboratories, although I like to think that they are, on the whole, at least as good. In all cases, however, my object has been to illustrate principles rather than give details of equipment.

The list of NPL colleagues, experts in their subject, who have helped with advice, suggestions and constructive criticism of sections of the book would be too long to include, but I wish to express my gratitude to them all, and to R. A. R. Tricker who read and criticized the whole manuscript.

In the choice of symbols, signs, abbreviations and graphical displays I have in general followed the recommendations of the Symbols committee of the Royal Society (1971). The Weights and Measures Act, 1963, uses the spelling ' kilogramme '. ' Kilogram ' is one of several abbreviations officially authorized for the purposes of trade. I have used it for brevity. It is the standard spelling in the USA and several other English-speaking countries.

P. VIGOUREUX

LIST OF SYMBOLS

For names and symbols of SI prefixes see table 2.4, page 11

A	ampere
B	magnetic flux density
C	coulomb
C	capacitance
c	speed of light
D	electric flux density
E	electric field strength, electromotive force
e	electronic charge
F	farad
F	force
f	force
g	gram
g	acceleration due to gravity
H	henry
H	magnetic field strength
h	Planck's constant
Hz	hertz
I	electric current
J	joule
J	electric current density
j	90° rotation operator (see note, page viii)
K	kelvin
k	Boltzmann's constant
kg	kilogram
L	self inductance, angular momentum, length
l	length, distance
M	mass, mutual inductance, magnetic moment
m	metre
m	mass
N	newton
n	rotational speed
Pa	pascal
Q	electric charge
$\tilde{R}$	resistance
r	distance, resistance
rad	radian

S	siemens
s	second
T	tesla
t	time, thickness
T	thermodynamic temperature
V	volt
V	electric tension, potential difference, electromotive force
v	velocity
W	watt
W	energy
Wb	weber
Z	impedance

γ	electric conductivity, gyromagnetic ratio
ϵ	electric permittivity of a medium
ϵ_0	electric constant, (permittivity of vacuum)
λ	wavelength
μ	magnetic permeability of a medium, refractive index of a medium, mass
μ_0	magnetic constant, (permeability of vacuum)
ρ	volume density of electric charge
Φ	magnetic flux, magnetic pole strength
Ω	ohm
ω	angular frequency

∇	operator called ' del ' or ' nabla '

$\nabla . \boldsymbol{V}$ divergence of vector $\boldsymbol{V}$ $\left(= \dfrac{\partial V_x}{\partial x} + \dfrac{\partial V_y}{\partial y} + \dfrac{\partial V_z}{\partial z} \right)$

$\nabla \times \boldsymbol{V}$ curl of vector $\boldsymbol{V}$ $\left(\text{vector with components } \dfrac{\partial V_z}{\partial y} - \dfrac{\partial V_y}{\partial z}, \ldots, \ldots \right)$

$\boldsymbol{P} \times \boldsymbol{Q}$ vector product of vectors $\boldsymbol{P}$ and $\boldsymbol{Q}$ (vector with components $P_y Q_z - P_z Q_y, \ldots, \ldots$)

$\dot{x}$ dx/dt

$|x|$ modulus of x

NOTE ON THE OPERATOR j

Addition of vectors can be simplified by the use of the operator j to rotate a vector 90° from some reference direction. The operator j^2 rotates the angle 180° and is thus equal to -1. The vector indicated by $a \cos \theta + j\, a \sin \theta$ has amplitude a, and components $a \cos \theta$ in the reference direction, and $a \sin \theta$ in the perpendicular direction; it is more briefly written $a\ e^{j\theta}$.

The resultant of vectors $a_1\ e^{j\theta_1}$, $a_2\ e^{j\theta_2}$, ..., is the vector $A\ e^{j\Theta}$, where $A \cos \Theta$ is equal to $a_1 \cos \theta_1 + a_2 \cos \theta_2 + \ldots$, and $A \sin \Theta$ is $a_1 \sin \theta_1 + a_2 \sin \theta_2 + \ldots$. Interest often lies not in the direction Θ of the resultant, but only in the square of the amplitude A. This quantity, A^2, is obtained by multiplying the resultant vector by its conjugate, i.e. by the resultant of $a_1\ e^{-j\theta_1} + a_2\ e^{-j\theta_2} + \ldots$; it is real.

We can also use the operator j to indicate phase. Thus at angular frequency $\omega\ (= 2\pi f)$ the $\pm 90°$ phase differences that distinguish inductive and capacitive reactance from resistance can be indicated by writing $j\omega L$ and either $1/j\omega C$ or $-j/\omega C$.

More generally an impedance $Z\ e^{j\phi}$ has magnitude Z and phase angle ϕ.

ABBREVIATIONS

Abbreviation	*Meaning*
a.c.	alternating current
BIPM	International Bureau of Weights and Measures
CGPM	General Conference of Weights and Measures
CGS	centimetre-gram-second system of units
CIPM	International Committee of Weights and Measures
d.c.	direct current
e.m.f.	electromotive force
MKSA	metre-kilogram-second-ampere system of units
NBS	National Bureau of Standards, USA
NPL	National Physical Laboratory, UK
NSL	National Standards Laboratory, Australia
p.d.	potential difference
RC	resistance-capacitance (as in RC network, etc.)
SI	International System of Units

To my mother
in remembrance

CONTENTS

CHAPTER 1

nature of measurement

1. *Object and scope*

MEASUREMENT is the process of comparing the quantity to be measured with another quantity of the same kind, or with quantities of other kinds, from which the magnitude of the first can be deduced. The mass of a body, for example, might be measured by comparison with the mass of a copy of the national standard kilogram. On the other hand a measurement of the speed of light might involve two separate measurements, one of time or frequency, and one of length, each of these two quantities being compared with the appropriate standard, say a clock for time, and a metre rod or a wave of a specified light for length.

In all cases standards are needed, but their quality and the ease or complexity of the process of comparison vary with the precision desired. The yard or metre sticks and the ' weights ' and scales handy and adequate in shop and market give place to optical interferometers on the one hand, and to accurate weighing masses and delicate balances on the other, when the precision required is a part or two in 100 million rather than in one thousand.

In the words of A. H. Cook, ' Measurement is the nervous system of technical life. Through measurement we learn of the physical world around us; through measurement we put that knowledge into numerical language, to which we apply mathematical methods to organize it into logical systems; through measurement this systematic knowledge is applied to change the physical world by engineering. Precise measurement is required for exact knowledge and economical design; convenient measurement is required for ready communication and efficient organization.'

The practice of measurement is almost as old as the human race, but the need for it has increased, and is increasing more and more rapidly with progress in technology, which not only demands higher and higher accuracy, but produces the facilities needed to meet this requirement.

The result of measurement is expressed as the ratio of the magnitude of the quantity measured, for example length, to a magnitude called the unit of the quantity. Thus the statement that the length of a beam is 6 metres is equivalent to saying that the beam is six times as long as the unit of length, or metre.

The unit is often, but not always, represented by a standard. The prototype of the kilogram for mass, and the period of a particular transi-

1

tion of the caesium atom for time, are examples. There are quantities, however, like energy, for which the unit is not represented by a standard, but whose magnitude is obtained from measurements of other quantities.

The scope of measurement ranges from the very crude to the highly accurate according to requirements, but here we shall be concerned more especially with the accurate measurements made to realize the units of various quantities, and with the standards which serve to maintain them.

The importance of measurement in scientific work was well understood by Clerk Maxwell, who, in the preface to his *Treatise on Electricity and Magnetism* wrote: ' The most important aspect of any phenomenon from a mathematical point of view is that of a measurable quantity. I shall therefore consider electrical phenomena chiefly with a view to their measurement, describing the methods of measurement, and defining the standards on which they depend.'

2. *Errors and uncertainties*

As discussions on realization and maintenance of units and on comparison of standards, to which most of this book is devoted, involve statements of accuracy or of errors, we give here an introduction to the subject of errors and uncertainties.

In the context of measurement, *accuracy* refers to nearness to the true value, *precision* to ability to detect small changes, *repeatability* to constancy of readings. *Uncertainty* refers to departure from the true value; although suggesting the reverse of accuracy, it is quoted in the same terms, viz. in parts in 10^3, in 10^6, etc., and it expresses the same notion, being perhaps a more logical term since, with the normal method of quoting, it increases with the number expressing it, whereas accuracy decreases as that number increases.

Reports of accurate measurements should provide sufficient information to enable the reader to examine critically the claims made for the accuracy of the result. A bald statement of the mean of a number of independent readings of the quantity being measured conveys no idea of the reliability of the result. A statement of the mean, accompanied by the probable uncertainty or the standard uncertainty, is more useful, but more valuable still is a table of individual results. Tables are however not as easy to study as graphical displays, and of these the most informative is one in which each result is represented by a point on the graph.

A satisfactory method of dealing with observational errors, often called ' random errors ', is shown in fig. 1.1, which displays 40 independent results obtained in the course of a determination of the gyromagnetic ratio of the proton. This plot is easy to prepare. First we decide on the precision of the measurements; in this case it was adjudged to be 1 in the seventh digit. The abscissa having been scaled accordingly, the individual results are read as they occur in the laboratory notebook, and points are marked at the corresponding abscissae, on the first horizontal

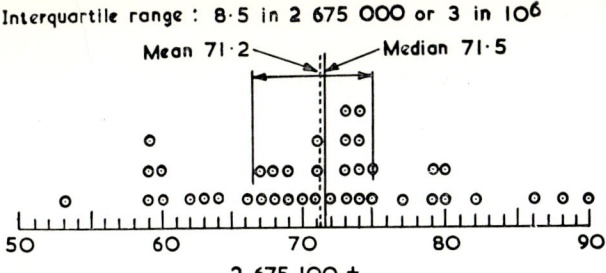

Fig. 1.1.   Display of 40 measurements of the gyromagnetic ratio of the proton.

line.   If the same result occurs twice, the second point is marked on the line above, and so on.   We end with a type of histogram, which for most experiments is roughly of ' gaussian ' shape, y proportional to $\exp(-x^2)$, if there is a large number of points.   Here 400 rather than 40 might have been needed to produce a better ' gaussian ' pattern, but the 40 were sufficient for the purpose of the investigation.

The easiest way to extract from this display the ' best ' value of the quantity being determined, and the degree of confidence it deserves, is to find the vertical line with an equal number of points, 20 here, on either side.   This operation is performed by mere counting, and the result, called the *median*, is here 71·5, whereas the *mean*, obtained rather less quickly by arithmetic, is 71·2.   That there is little to choose between the two is suggested by the confidence limits now to be determined. Vertical lines on each side of the median, with a quarter of the points between each and the median, are separated by an interval called the *interquartile range*, or sometimes ' 50% *zone* ', because there is a 50% chance of a single measurement of the quantity falling within this zone. Half the interquartile range is the *probable uncertainty* of a single measurement.   If the total number of points is n, the *probable error*, or *probable uncertainty*, of the mean is half the interquartile range divided by the square root of n; in this case it is 0·5 in 10^6.   As n increases, the chance that the mean differs from the mean of an infinite number of readings by less than the probable error of the mean tends to 50%.

Some authors prefer to use the *standard deviation* and the *standard error*, which are based on a 68·3% probability and are more convenient for mathematical work, although their calculation requires a good deal more arithmetic.   They are approximately equal to 1·5 times the probable uncertainty of a single measurement and the probable uncertainty of the mean respectively.   For measurements like the one given here as example, where the systematic error discussed below is much larger than the random error, the simple, quick method is adequate.

3

Systematic errors are those which affect each measurement to the same extent, and are not therefore reduced by a large number of readings. As however the number which should be taken to keep the random error low compared with the systematic error depends on the size of the latter, it is worth trying to estimate systematic errors when planning the experiment. This estimate is more uncertain than that of observational errors, which are determined by experiment, and as some systematic errors might never be suspected, they cannot be allowed for, although some of them can be avoided by varying as much as possible the routine of the measurements. Their estimate calls for ingenuity and pessimism, but here again it is all important that the reader should be told the details of the process.

Once these errors have been estimated, they must be combined. Measurements in natural science are as a rule subject to systematic errors from many, often independent, causes. The temperature of several components, the acceleration due to gravity, some weighing masses, atmospheric pressure, the dew point, the resistance of a resistor, the e.m.f. of a Weston cell, the diameter and length of coils, the magnetic susceptibility of the coil formers, might all enter the result of an investigation, and each one might be subject to a systematic error. But these errors would in general be uncorrelated; the calibration of thermometers, for instance, is independent of the magnetic susceptibility of the coil formers. For the purpose of combining these errors we are therefore justified in treating them as we would random errors. We therefore add their squares to the square of the observational error and take the square root of the sum as the total probable uncertainty of the investigation.

Further reading:

Barford, N. C. 1967. *Experimental Measurements: Precision, Error and Truth.* Addison-Wesley, London.

Topping, J. 1962. *Errors of Observation and their Treatment.* Chapman & Hall, London.

Vigoureux, P. 1966. ' Errors of observation and systematic errors '. *Contemp. Phys.*, **7**, 350–357.

4

UNTIL comparatively recently the system of units used for mechanics in the British Isles and most English-speaking countries was based on three independent units, the foot, the pound and the second. The actual standard of length was the yard of the Board of Trade, but a third of a yard, or foot, was in fact the unit for mechanical science, and gave rise to the unit of force called poundal, namely the force required to accelerate a mass of 1 pound at the rate of 1 foot per second per second. Like the yard, the pound was maintained by a material standard, a mass of platinum also kept at the Board of Trade, but the third ' base ' unit, the second, was deduced from astronomical measurements.

This system, although used in English-speaking countries well into the second half of the twentieth century for mechanical and civil engineering, was in the second half of the nineteenth century superseded for scientific work by one founded on the centimetre, gram and second as base units. The reason was that the electrical industry was rapidly increasing in importance and the practical electric units as we know them now, the ampere, volt, ohm, etc., had, by international agreement, been defined in terms of the centimetre, the gram and the second, and were used as much in English-speaking as in other countries.

There is no doubt that a single system of electric units in universal use, was a great help to progress in electrical science and engineering; nevertheless it had the disadvantage of not being coherent. A system of units is said to be coherent when derived units are obtainable from the base units without the use of a factor of proportionality other than unity. Speed, for instance, had as unit the centimetre per second; the unit of force, the dyne, was the gram-centimetre per second, per second, or $g \, cm \, s^{-2}$; and the unit of energy, the erg, was the dyne-centimetre. The units of the centimetre-gram-second (CGS) electromagnetic system form a coherent system of mechanical, electric and magnetic units, but the electric units of this system turn out to have inconvenient values: although the unit of electric current would be 10 times the present unit and perfectly acceptable, the units of electric potential and resistance would be absurdly small. By 1881, however, international agreement had been reached to fix the practical unit of potential, the volt, at 10^8 CGS units, which is of the same order as the e.m.f. of a primary cell, and the unit of resistance, the ohm, at 10^9 CGS units, which is of the order of the resistance of a column of mercury 1 m long and 1 mm² in cross-section. The unit of electric current, the ampere, was made a tenth of the CGS unit, so as to secure a coherent system of practical electric units.

Although these values were chosen to suit the needs of telegraphy, which was then a most important activity of electrical engineers, they happen to be satisfactory also for heavy electrical industry and for electronics.

The magnetic units were left at their CGS values, perhaps not surprisingly, for since the CGS unit, subsequently called ' gauss ', is of the order of the magnetic flux density of the Earth's magnetic field, and geomagnetism was at that time an important division of the science of magnetism, there seemed no point in changing the unit for one 10^4 times larger. Coherence was thereby lost to electromagnetism as it had already been lost to the system embracing the mechanical units and the practical electric units.

Whereas the electric units, by the agreement of 1881, were chosen to be of suitable magnitude for everyday use, and whereas the centimetre and the second are acceptable units of length and time, the gram however is, on the whole, too small for the general needs of man, which are better served by a unit nearer the magnitude of the pound or the kilogram. Moreover the CGS unit of force, the dyne, is too small to be felt by man unaided by instruments, and the unit of energy, the erg, is much too small; on the other hand the unit of energy provided by the practical electric units, the volt-ampere-second, or watt-second, called the joule, which equals 10^7 ergs, is of satisfactory size.

These considerations, the advantages of coherence, and the fortuitous circumstance that a mechanical system based on the metre and the kilogram has precisely the same unit of energy as is provided by the practical electric units, led G. Giorgi in 1902 to propose a system based on the metre, the kilogram, the second and one of the practical electric units. He moreover pointed out that if magnetic field strength were expressed as amperes per metre instead of the 4π times amperes per metre which is the definition corresponding to that of the CGS system, the number π would disappear from most electric and magnetic formulae involving rectilinear geometry, but would appear, as is natural, in those involving circles or spheres.

The choice of the metre and kilogram as base units is not only convenient in yielding suitable magnitudes for the units of force and energy, but also logical, since the prototypes on which the metric system is based are, or were at that time, those of the metre and the kilogram.

The International Electrotechnical Commission eventually chose the ampere as fourth base unit of the system, which became known as MKSA or Giorgi system, and in 1948 the 9th General Conference of Weights and Measures recommended it for science and technology, as well as for commerce and industry.

Units and standards of measurement for all those purposes are now largely agreed upon by most nations. The basis of this agreement is the

6

Metre Convention (1875) by which the adhering nations recognized the metric system, even though they did not all adopt it at once to the exclusion of other systems. In order to foster the use of the metric system and preserve uniformity of standards, they established the General Conference of Weights and Measures (CGPM) which convenes at intervals of about four years. The International Committee of Weights and Measures (CIPM), advised by a number of ' consultative committees ', prepares recommendations for submission to the CGPM, implements its decisions and supervises the International Bureau of Weights and Measures (BIPM), a central laboratory, at Sèvres near Paris, for intercomparison of standards.

In the Giorgi system the forces between electric charges and between magnetic poles in otherwise free space are given by

$$f = \frac{QQ'}{4\pi\epsilon_0 r^2} \quad (2.1), \qquad f = \frac{\Phi\Phi'}{4\pi\mu_0 r^2} \quad (2.2)$$

where ϵ_0 is called the electric constant, μ_0 the magnetic constant, f stands for mechanical force, r for distance, Q for electric charge and Φ for magnetic pole strength. The definition of the ampere, below, is

---

DEFINITIONS OF THE BASE UNITS OF SI

metre (m)
   The metre is the length equal to 1 650 763·73 wavelengths in vacuum of the radiation corresponding to the transition between the levels $2p_{10}$ and $5d_5$ of the krypton-86 atom.
kilogram (kg)
   The kilogram is the unit of mass; it is equal to the mass of the international prototype of the kilogram.
second (s)
   The second is the duration of 9 192 631 770 periods of the radiation corresponding to the transition between the two hyperfine levels of the ground state of the caesium-133 atom.
ampere (A)
   The ampere is that constant current which, if maintained in two straight parallel conductors of infinite length, of negligible circular cross-section, and placed 1 metre apart in vacuum, would produce between these conductors a force equal to 2×10^{-7} newton per metre of length.
kelvin (K)
   The kelvin, unit of thermodynamic temperature, is the fraction 1/273·16 of the thermodynamic temperature of the triple point of water.
candela (cd)
   The candela is the luminous intensity, in the perpendicular direction, of a surface of 1/600 000 square metre of a black body at the temperature of freezing platinum under a pressure of 101 325 newtons per square metre.
mole (mol)
   The mole is the amount of substance of a system which contains as many elementary entities as there are atoms in 0·012 kilogram of carbon 12.
   Note. When the mole is used, the elementary entities must be specified and may be atoms, molecules, ions, electrons, other particles, or specified groups of such particles.

Table 2.1

equivalent to ascribing to μ_0 the value $4\pi \times 10^{-7}$ henry per metre. The electric and magnetic constants are not independent, being connected, according to Maxwell's electromagnetic theory, by the equation

$$\mu_0 \epsilon_0 c^2 = 1 \tag{2.3}$$

where c is the speed of light in free space, a universal constant.

Although we shall not have occasion to apply the equations of electromagnetism known as Maxwell's equations, we give them here for completeness.

In free space, in the absence of electric charges and currents,

$$\nabla \times \boldsymbol{H} = \dot{\boldsymbol{D}} \quad (2.4), \qquad -\nabla \times \boldsymbol{E} = \dot{\boldsymbol{B}} \tag{2.5}$$

the dot indicates differentiation with respect to time, where the vectors H and E are magnetic and electric field strength, and where D and B, the electric and magnetic flux densities, are abbreviations for $\epsilon_0 E$ and $\mu_0 H$ respectively, and are subject to the conditions

$$\nabla \cdot \boldsymbol{D} = 0 \quad (2.6), \qquad \nabla \cdot \boldsymbol{B} = 0 \quad (2.7).$$

When we are concerned solely with free space it is often convenient to use only the two quantities E and B by substituting for D and H in (2.4)

DEFINITIONS OF SOME DERIVED UNITS

The unit of force, *newton* (N), is the force needed to accelerate a mass of 1 kg at the rate of 1 m/s².

The unit of energy, *joule* (J), is the work done by a force of 1 N acting over a distance of 1 m.

The unit of power, *watt* (W), is the power needed to perform work at the rate of 1 J/s.

The unit of potential difference and of electromotive force, *volt* (V), is the difference of electric potential between two points of a conducting wire carrying a constant current of 1 ampere, when the power dissipated between these points is equal to 1 watt.

The unit of electric resistance, *ohm* (Ω), is the electric resistance between two points of a conductor when a constant potential difference of 1 volt, applied between these two points, produces in the conductor a current of 1 ampere, the conductor not being the seat of any electromotive force.

The unit of quantity of electricity, *coulomb* (C), is the quantity of electricity transported in one second by a current of 1 ampere.

The unit of capacitance, *farad* (F), is the capacitance of a capacitor between the plates of which there appears a potential difference of 1 volt when it is charged by a quantity of electricity of 1 coulomb.

The unit of inductance, *henry* (H), is the inductance of a passive circuit in which an electromotive force of 1 volt is produced when the electric current in the circuit varies uniformly at the rate of 1 ampere per second.

The unit of magnetic flux, *weber* (Wb), is the magnetic flux which, linking a circuit of one turn, produces in it an electromotive force of 1 volt as it is reduced to zero at a uniform rate in 1 second.

Table 2.2

which then becomes

$$c^2 \nabla \times \boldsymbol{B} = \dot{\boldsymbol{E}} \qquad (2.8)$$

which, with (2.5), yields E and B in free space.

In continuous media, where there may be electric currents and electric charge, but not isolated magnetic poles, (2.5) remains unaltered but (2.6) and (2.4) are replaced by

$$\nabla . \boldsymbol{D} = \rho \quad (2.9), \qquad \nabla \times \boldsymbol{H} = \dot{\boldsymbol{D}} + \boldsymbol{J} \qquad (2.10)$$

where ρ is the volume density of electric charge, and J the vector density of electric current, connected with the electric field strength E by

$$\boldsymbol{J} = \gamma \boldsymbol{E} \qquad (2.11)$$

where γ is electric conductivity. Moreover the electric and magnetic constants ϵ_0 and μ_0 are replaced by the permittivity ϵ and the permeability μ of the medium, the ratios ϵ/ϵ_0 and μ/μ_0 being called relative permittivity and relative permeability.

Expression (2.1) shows why it is necessary to build the system from four, not three base units. The coherent unit of quantity of electricity derived from the metre, kilogram and second would be obtained by leaving out the factor $4\pi\epsilon_0$ from that expression; but then that unit, and the corresponding unit of current, would be of an inconvenient size and very different from the coulomb and ampere, which any system must retain if it is to be acceptable. The definition of the ampere, page 7, contains an arbitrary factor chosen to make it equal to the ' practical ' ampere, and it is this factor which turns it from a derived to a base unit.

Some of the units of the MKSA system had no names because they had not been widely used until the general adoption of the system. These units can be expressed in terms of others; the unit of force, for instance, would be the joule per metre; but as a short name is often convenient, General Conferences of Weights and Measures have from time to time recommended names as need has arisen. For example the name ' newton ' has been given to the unit of force; the unit of magnetic flux, the volt-second, has been named the ' weber ', and the unit of magnetic flux density, or weber per square metre, the ' tesla '.

The MKSA system of Giorgi admirably covers mechanics and electromagnetism, but it does not provide for other branches of science, heat for instance. In the hope of securing world-wide uniformity in the units employed in natural science, the 11th CGPM, in 1960, added to the units metre, kilogram, second and ampere, the kelvin for thermodynamic temperature, the candela for luminous intensity, and the radian and steradian for plane and solid angle. The first two joined the original four in being called ' base ' units, whereas the last two were called ' supplementary ' units. Any unit formed from two or more of those eight is called ' derived '. The MKSA system of units thus broadened is called the International System of Units, denoted by the initials SI of ' Système International ', and it is probably true to say that it is the most

9

satisfactory system of units we have had so far, in that it caters for the commercial and industrial activities of man, as well as for the needs of science.

When the 14th CGPM meets, in 1971, it will very probably add the mole, the unit of amount of substance, used in chemistry, to the list of base units, thus making them seven in all.

We should not attribute too deep a meaning to the words ' base ' and ' supplementary ' used in this context. The units of length, mass and time are, for the present at any rate, arbitrary and entirely independent of one another, and are thus properly called base units of a system founded on them. The ampere is defined in terms of the metre, kilogram, second, and of a factor chosen to make its value what it is: only in the arbitrary choice of the factor is it therefore comparable to the first three base units. The kelvin is independent of any other base unit and is

SUPPLEMENTARY AND DERIVED UNITS

Quantity	Unit	Symbol	
Supplementary Units			
Plane angle	radian	rad	
Solid angle	steradian	sr	
Derived Units			
Area	square metre	m^2	
Volume	cubic metre	m^3	
Frequency	hertz	Hz	$(1/s)$
Density	kilogram per cubic metre	kg/m^3	
Speed, velocity	metre per second	m/s	
Angular velocity	radian per second	rad/s	
Acceleration	metre per second squared	m/s^2	
Angular acceleration	radian per second squared	rad/s^2	
Force	newton	N	$(m.kg/s^2)$
Pressure	pascal	Pa	(N/m^2)
Viscosity (dynamic)	pascal second	Pa.s	
Viscosity (kinematic)	metre squared per second	m^2/s	
Work, energy, quantity of heat	joule	J	$(N.m)$
Power	watt	W	(J/s)
Quantity of electricity	coulomb	C	$(A.s)$
Electric tension, potential difference, electromotive force	volt	V	(W/A)
Electric field strength	volt per metre	V/m	
Electric resistance	ohm	Ω	(V/A)
Electric conductance	siemens	S	(A/V)
Capacitance	farad	F	$(A.s/V)$
Magnetic flux	weber	Wb	$(V.s)$
Inductance	henry	H	$(V.s/A)$
Magnetic flux density	tesla	T	(Wb/m^2)
Magnetic field strength	ampere per metre	A/m	
Magnetomotive force	ampere	A	

Table 2.3

arbitrary—the system could equally well have been based on the degree Fahrenheit had the latter been acceptable throughout the world. Alternatively, the thermodynamic scale might have been such as to make the mechanical equivalent of heat unity, but such a scale would have been unacceptable because of the inherent difficulty of determining it, and because the centigrade, or Celsius, scale from which the kelvin is derived, has been in use so much for so long. The choice of the radian, again, although natural and convenient, is arbitrary: the system could have been based on the right angle. In conclusion, there is no harm in the name ' base unit ' provided we attribute to it the meaning ' arbitrary ' rather than ' fundamental '.

The definitions of the base units and of some derived units, and a list of derived units with their names and symbols, are given in Tables 2.1, 2.2 and 2.3 below. Although we shall be concerned only with the first four base units, the kelvin and candela have been included in Table 2.1 for completeness, and the mole in anticipation of a very likely decision of the next CGPM. Table 2.4 is a list of names and symbols of prefixes which we may use instead of indicating the power of 10 by an index.

In the rest of this book the symbols listed in Tables 2.1 and 2.3 will be used for brevity, with the prefixes of Table 2.4 where appropriate.

NAMES AND SYMBOLS OF PREFIXES

Factor	Prefix	Symbol	Factor	Prefix	Symbol
10^{12}	tera	T	10^{-1}	deci	d
10^{9}	giga	G	10^{-2}	centi	c
10^{6}	mega	M	10^{-3}	milli	m
10^{3}	kilo	k	10^{-6}	micro	μ
10^{2}	hecto	h	10^{-9}	nano	n
10^{1}	deca	da	10^{-12}	pico	p
			10^{-15}	femto	f
			10^{-18}	atto	a

Table 2.4

There are some units which, although not included in the International System, are indispensable to man for his social, commercial and industrial activities and are extremely convenient for science. The hour, the day and the year, so long established and in such common use, will undoubtedly outlive the International System; the degree of angle is used extensively in scientific work, and there is no reason to force oneself to employ radians in cases when degrees are more convenient. SI is the servant of man, not his master.

Further reading:
The International System of Units. Translation of BIPM publication, HMSO, London, 1970.

CHAPTER 3
realization of base units

1. *The metre*

As the metre is now defined in terms of a wavelength of light, page 7, maintenance and intercomparison of standard 1 m bars no longer forms the basis of measurement, but has given place to the determination of length and distance by interferometry. In some applications it is convenient to use interferometry directly, but in industry and commerce when relatively low accuracy is satisfactory, it is often easier to use graduated measuring rods which have been calibrated by interferometric means.

Two basic techniques of interferometry are used for length measurement. One is a static system to determine the number of waves which fit between two partially reflecting surfaces, the other is a kinematic system using a counter to determine by how many wavelengths a reflector is displaced when it is moved, for example from one graduation of a scale to another. This system is readily adaptable to measurement in the factory for controlling, often by ' servo ' methods, the manufacture of components to very small tolerances.

One form of the static system is the Fabry-Pérot interferometer, whose essential parts are two parallel or nearly parallel surfaces facing each other, and coated with films which partially reflect and partially transmit light incident on them at normal, or nearly normal, incidence. A ray of light is reflected many times in succession at each surface, and at each reflection some light emerges by transmission through the surface. The emergent rays differ in phase and so form an interference pattern of ' fringes ' which move when one of the surfaces is slightly displaced. The phase differences and the pattern of fringes are calculated as follows:

In fig. 3.1 let MM, M′M′ be the films, and suppose for generality that the space in between has a refractive index μ. It might for instance contain a gas other than air, or air at a pressure different from that outside. A ray AB of monochromatic light, assumed for simplicity to have unit intensity, incident at a small angle to the normal to MM, is divided into two rays, one of intensity R reflected along BC, the other of intensity T transmitted along BE inclined at an angle θ to the normal. At E this ray is again split into EF of intensity T^2 and EB_1 of intensity RT. (Reflection and transmission coefficients are the same whether light enters or leaves the medium.) As θ will in practice be very small, there is no question here of total reflection, and at B_1 a ray of intensity

12

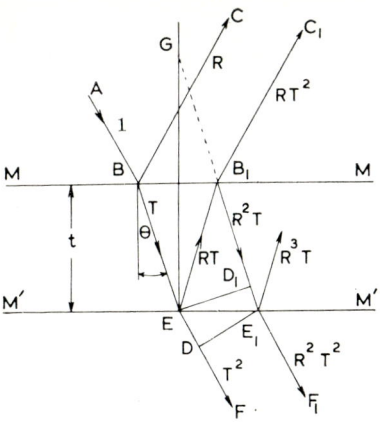

Fig. 3.1. Phase change of ray emerging after reflection from two parallel faces.

RT^2 leaves the plate in a direction B_1C_1 parallel to BC, and a ray B_1E_1 is reflected into the plate at B_1 to emerge as E_1F_1. The rays EF, E_1F_1 differ in phase. If ED_1, E_1D are the perpendiculars from E, E_1 to B_1E_1 and EF, the times light takes to travel ED in air and D_1E_1 in the medium are equal, so the effective path difference is EB_1D_1 in the medium. By the construction shown this is GD_1 or $2t \cos \theta$, which in air is equivalent to $2\mu t \cos \theta$, or to a phase difference ϕ given by

$$\phi = -4\pi\lambda^{-1}\mu t \cos \theta \qquad (3.1)$$

where λ is the wavelength in the air outside.

Figure 3.2 shows the successive emergent rays and their intensities. Their amplitudes are proportional to the square roots of the individual intensities. If the rays are brought to a focus, the intensity I is the square of the amplitude of the vector sum of the amplitudes T, RT, R^2T, . . . ,

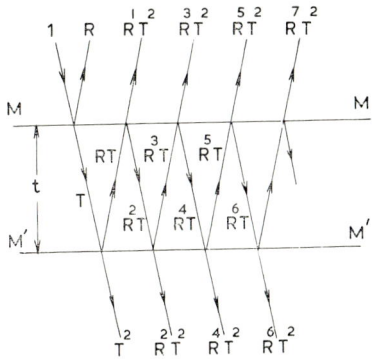

Fig. 3.2. Multiple reflections at two partially reflecting surfaces.

13

whose relative phases are 0, ϕ, 2ϕ, ... Proceeding as in the note, page viii, we multiply the vector sum by its conjugate and obtain the product of the two geometric progressions

$$I = T^2(1 + R\,e^{j\phi} + R^2\,e^{2j\phi} \ldots)(1 + R\,e^{-j\phi} + R^2\,e^{-2j\phi} \ldots)$$
$$= T^2(1 - R\,e^{j\phi})^{-1}(1 - R\,e^{-j\phi})^{-1}$$
$$= T^2(1 + R^2 - 2R\,\cos\phi)^{-1}$$
$$= T^2\{(1 - R)^2 + 4R\,\sin^2\phi/2\}^{-1}. \tag{3.2}$$

If we write F for $4R/(1 - R)^2$ the expression becomes

$$I = \left(\frac{T}{1 - R}\right)^2 \frac{1}{1 + F\,\sin^2\phi/2}. \tag{3.3}$$

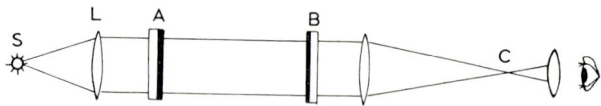

Fig. 3.3.   Fabry-Pérot interferometer.

In fig. 3.3 rays from the source S of monochromatic light are made parallel by the lens L, they travel through the glass plate A, and are partially reflected and partially transmitted by the film at A.   The transmitted rays are reflected in succession at B and at A, so that the beam of parallel light issuing from B and viewed by the telescope consists of all the rays of fig. 3.2.   Moreover, since the source of light is of finite size, there are rays brought to a focus off the axis, corresponding to small angles θ in (3.1), and because of axial symmetry, the interference pattern is circular and the fringes take the form of rings, as in fig. 3.4.   As the distance AB is increased the rings move outwards and new ones emerge from the centre.

In the Fabry-Pérot interferometer the reflection coefficient R is made 0·8 or more so that F of (3.3) is large.   Values of R as high as 0·95 can be obtained from films of silver evaporated on glass, but for most measurements the improved transmission of films of lower reflectivity is advantageous, and a value of about 0·8 for R is a good compromise.

Formulae (3.1) and (3.3) can be made to yield changes of θ corresponding to fringe separation and to fringe width at half intensity, which provide a figure of merit indicative of the precision with which the centre of a fringe can be located.

The phase difference between adjacent maxima of (3.3) is 2π, and as θ is very small, $\cos\theta$ may be replaced by $1 - \theta^2/2$ in (3.1).   Thus if adjacent maxima of (3.3) occur at θ_n and θ_{n+1}, (3.1) gives

$$2\pi = 4\pi\lambda^{-1}\mu t\theta_n(\theta_{n+1} - \theta_n). \tag{3.4}$$

14

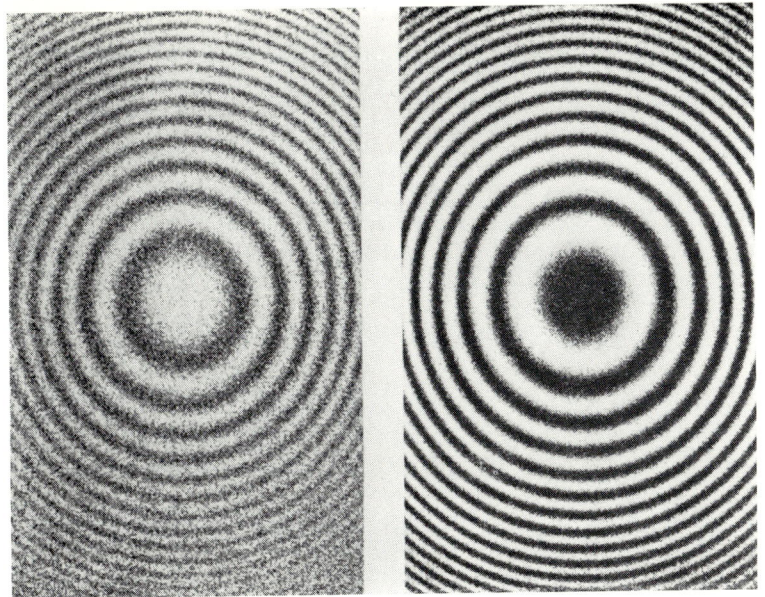

Fig. 3.4. Fabry-Pérot rings for krypton 86 (clear fringes) and for mercury 198 (less clear), obtained with a 40 cm path. (Crown copyright reserved.)

The half intensity angles are those for which $F \sin^2 \phi/2$ in (3.3) is equal to 1, and since F is large, $\sin^2 \phi/2$ is small, and may be replaced by $\phi^2/4$, so that the change of phase ϕ between a maximum and an adjacent half intensity point may be written $2F^{-1/2}$. Thus if θ_+ and θ_- are the half intensity angles on either side of θ_n, (3.1) gives

$$4F^{-1/2} = 4\pi\lambda^{-1}\mu t\theta_n(\theta_+ - \theta_-). \qquad (3.5)$$

Division of (3.4) by (3.5) leads to

$$\frac{\theta_{n+1} - \theta_n}{\theta_+ - \theta_-} = \frac{\pi F^{1/2}}{2}. \qquad (3.6)$$

If R is 0·8, F is 80, and the ratio of fringe separation to fringe width above is 14. If we compare this ratio with that for just two interfering beams, as in the Michelson interferometer, which even in the most favourable case of equal intensity of beams, is no more than 2, we understand why the multiple reflections of the Fabry-Pérot interferometer make it easier to locate the centre of the fringe.

We have hitherto supposed the radiation strictly monochromatic, in other words of a single frequency. Such is not the case even with good spectral lines, for so-called monochromatic radiation has a spectrum of frequencies for several reasons, of which we shall only mention two:

15

Doppler broadening and isotopic broadening. The former is due to the thermal motion of the atoms in the electric discharge tube producing the light. The atoms move randomly in all directions, so that some of them move predominantly towards the interferometer, others away from it; the former appear to radiate at a frequency relatively higher than the emitted frequency, by v/c, where c is the speed of light and v is the velocity of the atom in the direction of the light, whereas the latter seem to radiate at a correspondingly lower frequency. If the atoms have a Maxwell distribution of velocities, the curve of intensity against wavelength has the shape

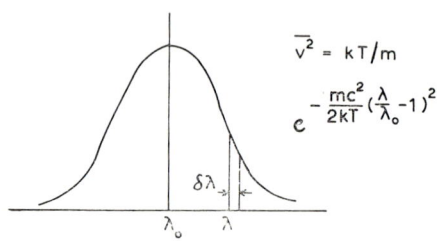

$$\overline{v^2} = kT/m$$

$$e^{-\frac{mc^2}{2kT}(\frac{\lambda}{\lambda_0}-1)^2}$$

Fig. 3.5. ' Doppler ' broadening caused by motion of atoms. The number of atoms radiating at wavelengths between λ and $\lambda + \delta\lambda$ is proportional to the ordinate at λ.

of fig. 3.5. The fringes produced by the radiation at the wavelength λ occur at slightly greater intervals than those formed by the central radiation λ_0. For small values of the spacing t of the plates, and therefore of the number or ' order ' n of fringes, the effect is of little consequence, but, since $2t$ is equal to $n\lambda_0$, when n reaches the value $\lambda_0/2(\lambda - \lambda_0)$, and therefore t the value $\lambda_0^2/4(\lambda - \lambda_0)$, the subsidiary fringes occur half way between the main ones, and as the spectrum contains not just those two wavelengths but a continuous distribution, the fringe pattern becomes blurred long before that separation is reached.

If the gas or vapour providing the source of light contains isotopes, i.e. atoms of the same element, but differing in atomic mass, the lines emitted by the different isotopes are not in general exactly coincident, and thus act like a comparatively broad line to yield as before an indistinct pattern at large plate separations. Furthermore isotopes of odd mass number often give broad spectral lines because each line is split into a number of very close components.

Before isotopes could be separated, the cadmium red line of 643·8 nm was one of the narrowest available. Examination of light from single isotopes subsequently showed that cadmium 114, krypton 86 and mercury 198 gave very narrow lines, especially when Doppler broadening was reduced to a minimum. The choice for the definition of the metre fell on the krypton-86 orange line at 606 nm because this line was found to be particularly reproducible in wavelength. Its source is operated at the triple point of nitrogen, 63 K, so that Doppler broadening, being

16

proportional to $(T/m)^{1/2}$, fig. 3.5, is also very small at that temperature.

The metre is defined by an atomic transition—unperturbed, i.e. in the absence of disturbance due to the causes mentioned above, or to any others. One problem in realizing the metre is to find a lamp reproducing the wavelength without serious disturbance. In practice the lamps do affect the wavelength to the extent of at least 1 or 2 in 10^8, depending upon operating conditions. Studies on these lamps have shown that under certain conditions some of the effects cancel, making it possible to get light of a wavelength corresponding very closely to that of the ideal unperturbed transition. These conditions are formulated in a recommendation approved by the CIPM but separate from the actual definition of the metre.

The krypton-86 line gives satisfactory fringes at path differences less than half a metre, but stabilized lasers, with much narrower lines still, have recently increased the workable distance to over 100 m. It is of course necessary to know the wavelength of the light used in terms of that of the krypton-86 line, and the relation for various useful optical and laser lines has been established by careful measurements.

In the use of the Fabry–Pérot interferometer to measure a fixed distance, the fraction of a fringe separation is estimated by means of the cross-wire and micrometer of the telescope, but the integral order is unknown. In measurements of high precision, however, the distance may already be known to a few wavelengths; if so, the method of ' excess fractions ' will yield the actual integer, but it entails knowledge of the excess fractions for 3 or 4 wavelengths. This information is often provided by a source emitting the several waves, the fringe patterns being separated by a spectroscope whose slit is placed at C, fig. 3.3. Figure 3.6 shows the fringes for four lines of cadmium.

Let the observed fractional orders for the four wavelengths be f_1, f_2, f_3 and f_4, and suppose the distance t is known to a few wavelengths of λ_1. If n_1 is the integer nearest to $2t/\lambda_1$, the actual distance will be given by

$$2t = (n_1 + x + f_1)\lambda_1 \qquad (3.7)$$

where x is a small positive or negative integer. The orders for the other lines can now be calculated by dividing the above by λ_2, λ_3, λ_4 for a few values of x, say . . . -3, -2, -1, 0, 1, 2, 3 . . . . The fractional parts ϕ, ψ, χ of the numbers thus obtained will vary with x, and for one value only, say x_1, will they be equal to the observed fractions f_2, f_3, f_4. The correct order is thus $n_1 + x_1 + f_1$ and the distance t is obtained to a high accuracy by multiplying this number by $\lambda_1/2$.

In one experiment with the cadmium red, green, blue and violet lines, whose wavelengths in air were taken as 643·8472, 508·5824, 479·9911 and 467·8152 nm, Benoît observed the fractional orders 0·82, 0·00, 0·79 and 0·93 with a probable uncertainty of perhaps 0·03 of a fringe. The interface distance, measured with a good micrometer, lay between 2·5

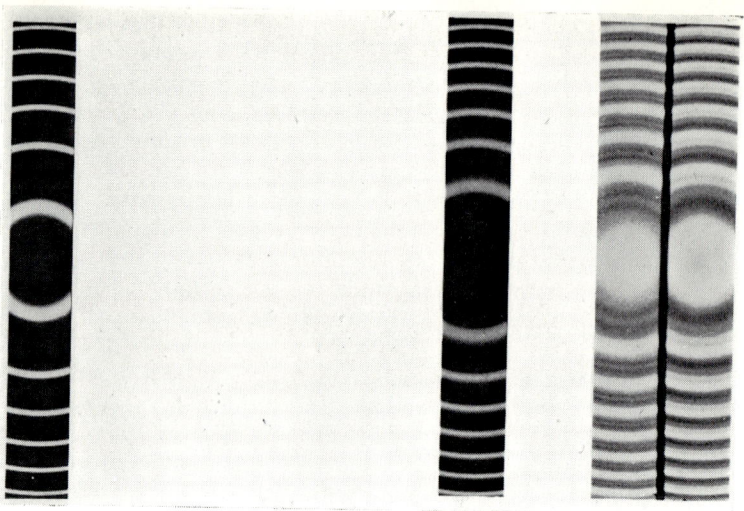

Fig. 3.6. Fringes in 4 visible radiations of cadmium. (Crown copyright reserved.)

and 3 μm below the nominal 10 mm. We take the mean and divide $2(10 - 2 \cdot 75 \times 10^{-3})10^6$ by 643·8472 to obtain 31 055 as the best guess for the order. Proceeding as explained above we calculate the orders of the green, blue and violet fringes for integral values of x from -4 to $+4$ to obtain Table 3.1. Looking in this table for the row, or rows, in which the fractional orders are, within experimental error, equal to the observed decimals 0·82, 0·00, 0·79 and 0·93, we find that for one value only of x, viz. -1, are these figures nearly reproduced. The order for the red line was thus 31 053·82 and the interface distance, product of this number by $\lambda_1/2$, is 9 996 958 nm.

Indeed we might go further and, after averaging the distance found in this way for the four lines, use the mean to calculate the wavelengths for better consistency, or alternatively, if it were thought that one wavelength, for instance that of the violet line, was subject to a greater uncertainty than the others, use the mean distance from the first three colours not only to re-calculate their wavelengths, but also to obtain a better figure for the wavelength of the fourth.

In this age of computers the method of excess fractions can be applied to cases when the uncertainty in the distance measured is as large as 1 mm or so. It is only necessary to extend the table both ways for larger values of x, and it helps to use more wavelengths, say 6 instead of the 4 of the table.

Turning now to the kinematic systems, what is required in many applications is the displacement of some component, to which one plate

18

x	red	green	blue	violet
−4	31050·82	39309·23	41650·74	42734·79
−3	1·82	10·50	2·08	6·16
−2	2·82	11·76	3·42	7·54
−1	3·82	13·03	4·76	8·92
0	4·82	14·30	6·10	40·29
1	5·82	15·56	7·44	1·67
2	6·82	16·83	8·79	3·05
3	7·82	18·09	60·13	4·42
4	8·82	19·36	1·47	5·80

Table 3.1. Determination of order of interference by method of excess fractions.

of the interferometer is rigidly fixed.   This determination is in many ways simpler than that of a fixed distance, because a photoelectric detector and counter can serve to count the number of fringes moving past the detector as the plate travels.   In conjunction with the photo-electric detector, a diaphragm in the image plane of the telescope excludes light except from the centre of the fringe system.

With adequate safety precautions a helium-neon laser can serve as source of light; with a carriage speed of, say, 2 mm/s there are about 6452 counts per second, and a displacement of 1 m corresponds to approximately 1 613 000 counts.   Here, as in all applications, the tabulated vacuum wavelength must be corrected for refractive index if the measurements are made in air.

In spite of the facilities which photoelectric detectors, counters and lasers provide for ' on the job ' measurement of length by interferometry, graduated line standards are still needed in industry, and their calibration is an important function of a standards laboratory.

In a machine built for the purpose at NPL the scale S, fig. 3.7, rests

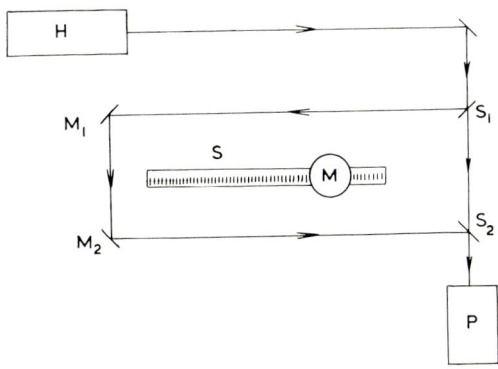

Fig. 3.7.   Automatic scale measurement by laser.   (Rowley and Stanley.)

on a carriage which can be translated by a hydraulic drive at a speed of 2 mm/s in the direction of the length of the scale. Light from a helium-neon laser H, parallel to and behind the bed of the machine, is directed by a system of mirrors to the fixed semi-reflectors S_1 and S_2. Part of the light goes straight through them to reach the photoelectric detector P, whilst part is reflected parallel to the scale and back parallel to itself by the mirrors M_1, M_2 mounted on the carriage, to reach S_2 where it joins the other beam and interferes with it. As the carriage moves, the length of the optical path changes and the intensity of the recombined beam varies sinusoidally, each period corresponding to a carriage movement of half a wavelength, 0·31 μm approximately. These intensity variations, or fringes, produce corresponding changes in the output of the photo-detector and work a counter. Above the scale, but fixed to the bed of the machine, a photoelectric microscope M with a slit parallel to the scale divisions detects light reflected from the illuminated scale. As a division moves past the slit, a signal derived from its passage opens the 'gate' to a counter, which in consequence becomes active, and starts counting the fringes detected by P. The next scale line stops this counter and opens the gate of a second counter which counts the number of fringes between this line and the next, and so on, the two counters being used alternately. As the counters provide a time scale and the speed of the carriage is reasonably constant, it is possible to interpolate, to a fraction of a fringe, the position of each scale line by measuring the time interval between the signal from the scale and the instant of the next fringe count. The number of fringes and the time intervals in micro-seconds representing the fractions are automatically punched on paper tape subsequently processed by a digital computer which prints the errors of the lines of the scale.

Further reading:

Barrell, H. 1962. 'The metre'. *Contemp. Phys.*, **3**, 415–434.
Rowley, W. R. C. and Stanley, V. W. 1965. 'The laser applied to automatic scale measurement'. *Machine Shop*, **26**, 430–432.
Tolansky, S. 1955. *An Introduction to Interferometry*. Longmans Green, London.

2. The kilogram

The kilogram is the mass of the prototype of the kilogram, and no realization is involved. There are however continual calls for compari-son of masses nominally of 1 kg with the prototype, or much more frequently with copies kept by various countries. The instrument for comparing masses of that sort is invariably a balance, the principle of which is as follows:

In fig. 3.8 let K, K' be the vertices of the cross-sections of two wedges

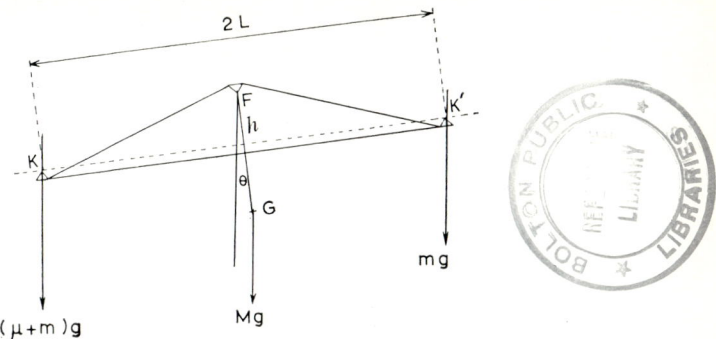

Fig. 3.8. Principle of balance.

or 'knives' of hard metal or semi-precious stone like agate or sapphire, fixed to a rigid frame carrying at F, called the fulcrum, a similar but inverted knife. This knife rests on a fixed horizontal plate, and K, K' support plates from which hang scale pans to carry the masses to be compared. Denote these masses, assumed equal, by m, and suppose that, in order to measure sensitivity, an extra small mass μ is added to one pan. Suppose the beam itself has mass M, denote the distance KK' by $2L$, call a the height of F above the line KK', and h the distance from F of the centre of gravity G of the beam. In the position of equilibrium FG will be inclined to the vertical at a small angle θ given by

$$g(m+\mu)(L-a \tan \theta) \cos \theta = gm(L+a \tan \theta) \cos \theta + gMh \sin \theta \quad (3.8)$$

and since the calibrating mass μ is always chosen small to keep θ very small,

$$\theta = \frac{\mu L}{Mh + 2ma} \quad (3.9)$$

approximately.

We conclude that if the sensitivity θ/μ is to be independent of the 'load' m, a must be zero. Every attempt is therefore made to keep the lines of the three knives in one plane, and of course parallel, independence of sensitivity on load being the criterion of correct adjustment.

If μ is removed and the beam is allowed to oscillate, its angular position ϕ at time t is given by the equation

$$I \frac{d^2\phi}{dt^2} + R \frac{d\phi}{dt} + (Mh + 2ma)g\phi = 0 \quad (3.10)$$

in which I is the moment of inertia of the system and R the resistance to motion for unit angular velocity. If θ is the initial value of ϕ, given by (3.9), the solution is

$$\phi = \theta \, e^{-(R/2I)t} \cos\left(\frac{Mh + 2ma}{I} - \frac{R^2}{4I^2}\right)^{1/2} t \quad (3.11)$$

21

indicating sinusoidal damped oscillations of period T given by

$$\frac{4\pi^2}{T^2} = \frac{Mh+2ma}{I}g - \frac{R^2}{4I^2} \qquad (3.12)$$

the second term on the right being always considerably smaller than the first, and almost always negligible.

The ratio of the sensitivity to the square of the period is then from (3.9) and (3.12)

$$\frac{\theta}{\mu T^2} = \frac{gL}{4\pi^2 I}. \qquad (3.13)$$

In good balances, with a equal to zero or nearly so, the sensitivity is from (3.9)

$$\frac{\theta}{\mu} = \frac{L}{Mh}. \qquad (3.14)$$

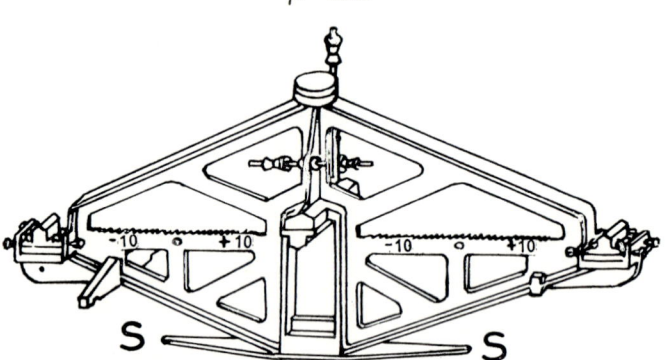

Fig. 3.9.   Beam of kilogram balance of the National Physical Laboratory.
(Crown copyright reserved.)

Balances are usually provided with three adjusting masses in the form of nuts movable along threaded rods, as in fig. 3.9, which shows the beam of a balance built at NPL for comparing masses of 1 kg.   Two of the nuts serve to adjust the rest point, i.e. to make θ zero or small in the normal position of equilibrium, and the third, on the vertical rod, to alter the sensitivity.   The beam is also fitted with a mirror which is part of an optical system to project on a scale 6 m away the image of an illuminated cross-wire.   As the beam swings, the image travels to-and-fro on the scale with a slightly damped harmonic motion.   The position of equilibrium, or rest point r, is found by noting successive extreme positions a_1, b_2, a_3, two at one end and one at the other; it is given by

$$r = \tfrac{1}{2}\left(\frac{a_1+a_3}{2} + b_2\right). \qquad (3.15)$$

Five positions are sometimes read for greater precision.

To determine the sensitivity two rest points r_1, r_2 are taken, the first

22

with a small known mass μ_1 added at one end of the beam, the other with a nominally equal mass μ_2 added at the other end. The sensitivity s is given by

$$s = \frac{|r_2 - r_1|}{\mu_1 + \mu_2}. \tag{3.16}$$

This balance has a rather elaborate suspension consisting, on each side, of two crossed knives and a cone, designed to ensure that the load is located precisely on each end knife of the beam. The suspension is illustrated diagrammatically in figure 3.10, which also shows a standard of mass resting on an auxiliary pan. A 'transporter' mechanism lifts the auxiliary pans, each with its weighing mass, and interchanges them on the scale pans. From the difference of rest point, and a prior or subsequent measurement of sensitivity, the difference between the masses of the standards plus their auxiliary pans is obtained. The difference between the masses of the pans alone can be determined by a similar procedure and allowed for.

Although the beam and the stirrups hanging from the several knives are lifted by an 'arrester' mechanism when the balance is not in use, the interchange of masses described above is effected without the beam being arrested, but only prevented from swinging. For this purpose two agate points carried by an independent mechanism below the beam are raised so as just to meet the underface of the flexible metal strip S S, fig. 3.9, fixed horizontally to the lowest part of the beam, which is thus brought temporarily to rest without disturbing the contacts between knives and planes.

The balance is mounted on a pillar in a vault near the centre of a large basement. The natural distribution of temperature in the vault being particularly steady and uniform, for the walls are thick, dispenses with the need for automatic thermostatic control, which for accurate balances often creates more problems than it solves. The arrestment and other controls are operated from outside the vault to avoid possible disturbance due to proximity of an observer.

As a result of these and other refinements, this balance can serve to compare kilogram masses to 0·001 mg, i.e. to 1 in 10^9, but the procedure needed to attain this accuracy occupies two days. The evening before the first day the bodies whose masses are to be compared are placed centrally on the interchangeable pans and the two sets of pan and mass are equalized to within 0·05 mg by placing appropriate small masses on the pans.

In the morning of the following day the observer, from outside the balance vault, brings the beam into a stable condition by releasing it and allowing it to swing for a few minutes. It is then steadied for a moment, but not arrested, next released, and after a swing or two a rest point r_1 is taken. The beam is steadied anew, the pan plates with their loads are interchanged, the beam is released, and the new rest point r_2 is deter-

23

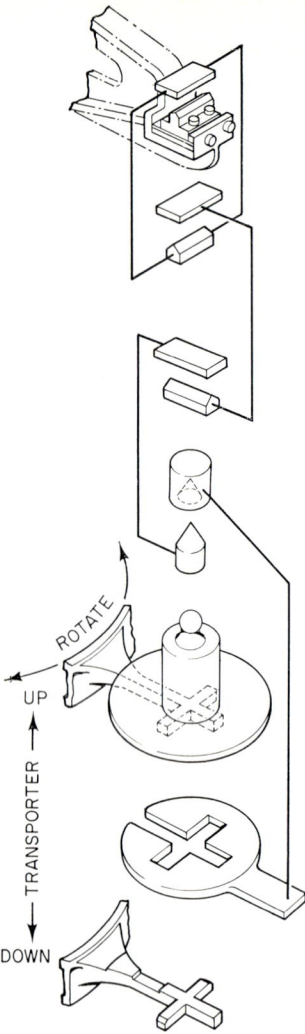

ROTATE

UP

TRANSPORTER

DOWN

Fig. 3.10. Suspension of scale pan of kilogram balance of the National Physical Laboratory. (Crown copyright reserved.)

mined as before. The procedure is continued until, say, five rest points have been obtained with the masses in one position, four in the other. Any drift of rest point from thermal or mechanical causes is thus minimized, and the difference Δr between the rest points for the two loadings is given by

$$4\Delta r = \tfrac{1}{2}r_1 + r_3 + r_5 + r_7 + \tfrac{1}{2}r_9 - r_2 - r_4 - r_6 - r_8. \qquad (3.17)$$

The sensitivity is measured, and nine more weighings made as before.

At the end of the day the vault is entered, the balance case is opened, the masses are interchanged but not the pans, and the next day the previous day's operations are repeated. If the masses are M_1, M_2 and the masses of the pans are P_1, P_2, the observations of one day yield

$$M_1 + P_1 - M_2 - P_2$$

and those of the other day $M_1 + P_2 - M_2 - P_1$; thus $M_1 - M_2$ is obtained by addition and, incidentally, $P_1 - P_2$ by subtraction.

This lengthy procedure is followed only when an accuracy of 1 in 10^9 is needed. In many cases the desired accuracy might be only 1 in 10^6, when a less elaborate procedure and a simpler and cheaper balance would be adequate.

Copies of the prototype of the kilogram can be compared with such high accuracy, and they exhibit such constancy, that it has not been possible to suggest an alternative unit of mass, based like the metre on some constant, which could be realized even with an uncertainty 100 times that of comparison and maintenance of kilogram masses, let alone with comparable uncertainty. The unit of mass is thus unlikely to be redefined for some time to come.

Further reading:

Gould, F. A. 1949. ' A knife-edge balance for weighings of the highest accuracy ', *Proc. Phys. Soc.* B, **62**, 817–828.

3. *The second*

Experience with atomic beams has shown that the period of transition of the hyperfine levels of caesium atoms provides a convenient and accurate standard of time interval, and we saw, page 7, that the second is now defined in terms of that period, the number of periods to make up one second having been chosen to secure as close agreement as possible with the former, astronomical, definition. Realization of the unit therefore entails the use of some device to compare that frequency with one which can be conveniently subdivided to produce ' seconds '. It does not follow that this procedure is also the most practical one for maintaining the unit from day to day, for if other transitions proved more convenient it might be preferable to compare once for all their frequencies with that of caesium, and use them instead for the daily routine of maintenance of the second. It happens however that although other transitions are available, the ' atomic clock ' using the caesium transition is practical and convenient, as might be expected, since convenience was one of the properties considered in choosing it to define the unit.

The frequency f of the radiation emitted or absorbed when an atom undergoes an energy change W is

$$f = \frac{W}{h} \tag{3.18}$$

where W is the change of energy and h is Planck's constant. The frequency of the so-called ' hyperfine ' transition of the caesium atom is approximately 9·2 GHz, which is in the centimetre range of wavelength, for which techniques of subdivision are well established. The problem is to produce transitions by application of microwave power and observe the frequency for which the rate of transition is a maximum.

Atoms of caesium, and of other alkali metals, have a single valency electron, whose spin may be in the same direction as the spin of the nucleus or in the opposite direction, and it is this difference which accounts for the difference of energy W above. The magnetic moments of the atoms in the two states are opposite and nearly equal, so that if a beam of atoms whose spins are maintained perpendicular to the direction of travel, and in a single plane, by a small uniform magnetic field in that plane, is made to cross a short region of large magnetic gradient in the same plane, the atoms are deflected in the plane in one direction or the other according to the sense of their electronic spin. If there are two similar regions of magnetic gradient, fig. 3.11, with a comparatively long region of the small magnetic field in between, an atom deflected by the first deflecting magnet is deflected further out by the second unless it has undergone a transition during its flight from one to the other, in which case it is deflected back towards the axis of the beam and reaches the detector.

Transitions are produced in the intermediate region by the radio-frequency field applied by the two cavity resonators. The number of atoms undergoing transitions is greatest when the radio frequency at which the cavities are excited is exactly given by (3.18); the sharpness of resonance is inversely proportional to the separation of the cavities.

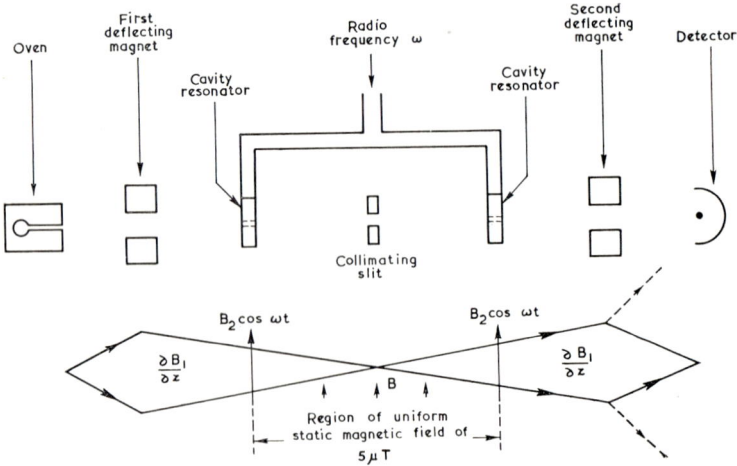

Fig. 3.11.   Principle of detection of resonance of caesium atomic beam.

26

Atoms stream through minute holes in the oven with a thermal velocity corresponding to about 370 K, and those which reach the detector, a hot wire of tungsten or of platinum-iridium, are ionized by it and drawn to a collector plate as an electric current, which is amplified to operate a meter or recorder.

We have hitherto mentioned only two energy states.   In spectroscopy the spin of the electron is denoted by J and is 1/2, and the spin of the nucleus, denoted by I, is 7/2 for caesium 133.   The levels resulting from the interaction of the spins, denoted by F, are $I \pm J$, i.e. 4 or 3, fig. 3.12. In the absence of a magnetic field the atoms would have energies corresponding to one or the other of these hyperfine levels, but the uniform magnetic field perpendicular to the beam splits the levels into 8 and 6 additional lines respectively as in fig. 3.12, and the energies of the Zeeman components F, m_F are proportional to the field, as indicated diagrammatically by their slope in the figure, unlike the central components which are almost independent of it, at any rate for low fields.

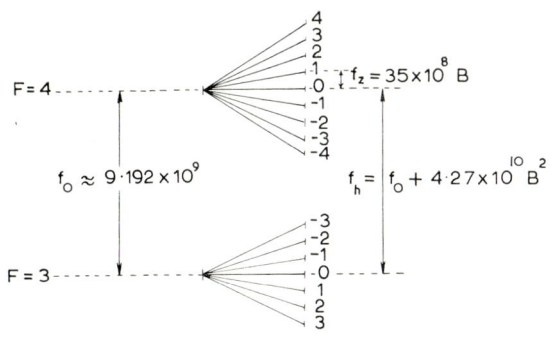

Fig. 3.12.   Hyperfine and Zeeman energy levels of caesium.

The frequency of the hyperfine transition corresponding to the central levels is

$$f = f_0 + 4 \cdot 27 \times 10^{10} B^2 \tag{3.19}$$

where f_0 is the frequency 9 192 631 770 Hz of the definition of the second, page 7, and B the magnetic flux density in tesla.   It might be thought therefore that the simplest procedure would be to eliminate the magnetic field completely.   As it would however be well-nigh impossible to do so, transitions would still occur at a number of frequencies, which would be so close together that it would be difficult to isolate the desired one, so a small magnetic field, large enough to separate the Zeeman components and avoid confusion between the desired transition and those involving Zeeman levels, is retained in the region between the cavity resonators and, if necessary, a correction is applied.   In practice the field is reduced to about one tenth of the Earth's field, say 5 μT, and the correction is just over 1 Hz.

The cavity resonators must be supplied with a frequency close to nominal and adjustable over a few hundred hertz, so that the resonance curve may be traced; that frequency is best derived by synthesis, an electronic process involving multiplication, division and addition, from an oscillator of some integral number of megahertz, to provide also lower frequencies by division, and eventually seconds.

CAVITY SEPARATION 2·6 m

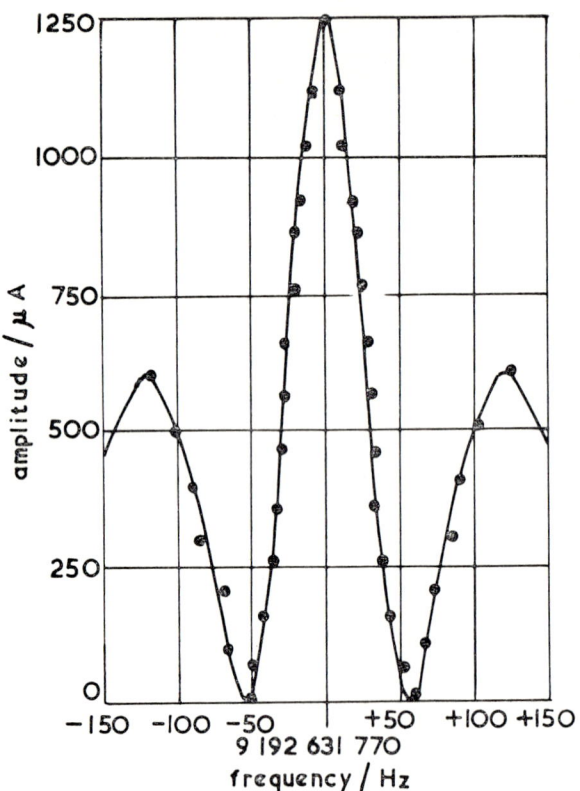

Fig. 3.13. Resonance curve of the long caesium beam of the National Physical Laboratory. (Crown copyright reserved.)

The resonance curve, fig. 3.13, of the long caesium beam of the National Physical Laboratory is very sharp, 50 Hz wide at half intensity, allowing the central frequency to be determined to about 0·1 Hz or 1 in 10^{11}. The precision of that standard, and of other atomic clocks, has made it possible to measure the variations of the rate of rotation of the Earth, fig. 3.14.

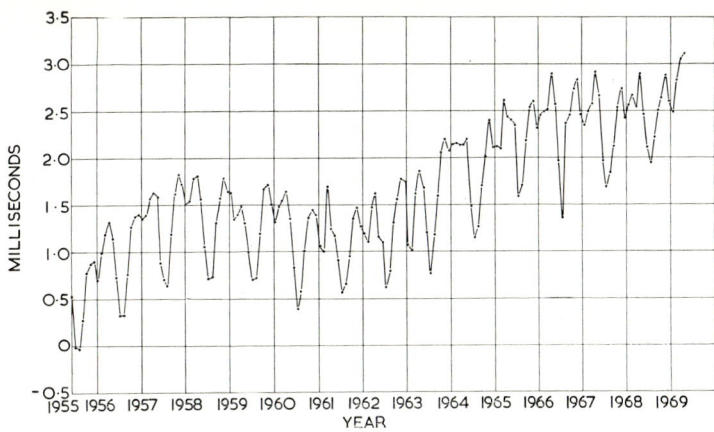

Fig. 3.14. Variation in the length of the day in terms of the atomic unit of time. (Crown copyright reserved.)

For many applications, however, a resonator, which needs to be excited by a generator of appropriate frequency, is not as convenient as an oscillator supplying power continuously at the frequency of the atomic transition. The NPL standard, and others more compact for field as well as laboratory use, have therefore been made to work in a servo loop. In principle a generator, for example a quartz crystal oscillator, whose frequency can be varied slightly by application of a potential difference to one of its components, controls as before a synthesizer which excites the cavity resonators of the caesium beam at a frequency close to that of transition, but based on the generator frequency. The function of the apparatus is then to produce a correction signal dependent on the small difference between the applied frequency and that of transition, and to apply this signal to the generator to alter its frequency until it corresponds to the frequency of transition.

In one caesium-beam frequency standard on the market this ' servo ' control is effected approximately as follows: a 5 MHz quartz oscillator, fig. 3.15, whose frequency is controllable over a small range by a d.c. voltage applied to a voltage-sensitive capacitor (a silicon diode), is phase modulated at a repetition frequency of 137 Hz, and successive frequency multiplication by 18 and by 102 bring the frequency to 9180 MHz. This frequency is not close enough to the transition frequency 9 192 631 770 Hz, so the 5 MHz oscillator is also made to operate a synthesizer. This instrument can provide almost any frequency, below a certain maximum, based on the 5 MHz input, and the frequency selected for the present purpose is 12·631 770 MHz plus the correction for the magnetic field in the particular caesium tube. This frequency is added to the 9180 MHz to give 9192·631... MHz. The output, on

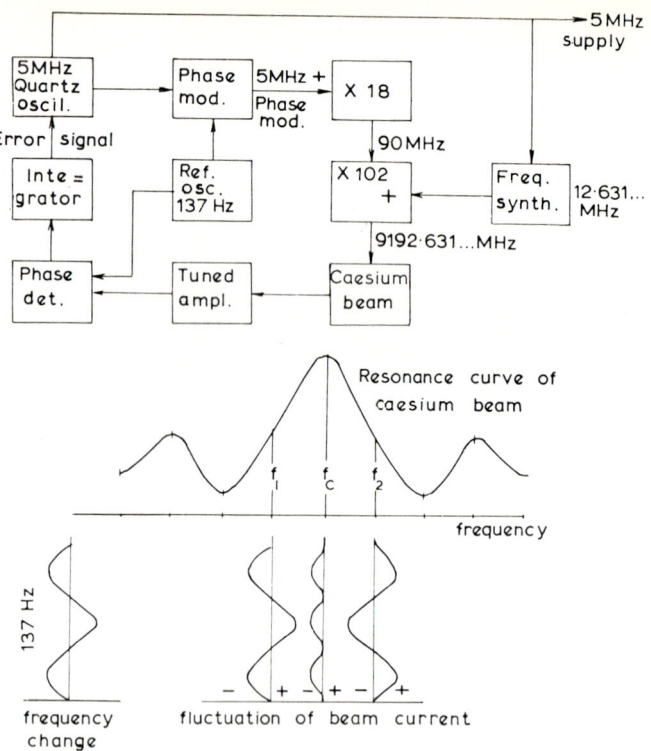

Fig. 3.15. Principle of servo control of quartz crystal at 5 MHz by caesium beam. (Hewlett-Packard.)

account of the phase modulation applied to the 5 MHz, is frequency modulated. Depending on whether the quartz oscillator gives exactly the nominal 5 MHz or not, the output current of the caesium beam is amplitude modulated at 274 Hz only or also at 137 Hz, for as shown in fig. 3.15, if the applied frequency is centred on f_1 the fluctuation of beam current is in phase with the frequency modulation, if it is centred on f_2 it is in antiphase, whereas if it is centred precisely on the frequency f_c of the peak of the resonance curve, the beam current fluctuates at twice the repetition frequency, and the fundamental is absent. The output current acts on a tuned 137 Hz amplifier followed by a phase detector, whose output is thus positive, zero or negative according to the value of the central frequency of the frequency-modulated microwave applied to the cavities. The error signal, integrated to remove the 274 Hz fluctuation and applied to the voltage sensitive component of the quartz oscillator, keeps the 5 MHz frequency almost exactly nominal, with only the slightest phase fluctuation.

30

It is not necessary for organizations other than standards laboratories, observatories and a few others of highly specialized character to own caesium beams or similar standards with which to realize the second, for many countries broadcast frequency and time signals by which any laboratory having a radio receiver can calibrate its wavemeters or electronic counters. These transmissions are in general accurate to 1 in 10^{11}, and the low-frequency carriers like Rugby 16 kHz (GBR) and USA 18 kHz (NBA) are not greatly affected by the movements in the ionosphere which cause Doppler shifts in the megahertz range, except in the small areas where the ground wave predominates.

Further reading:
Essen, L. 1948. ' The measurement of time '. *Vistas in Astronomy*, **11**, 45–67.
Hewlett-Packard. 1965. *Model 5060A Caesium Beam Frequency Standard Training Manual.* Hewlett-Packard Company, Palo Alto, Calif., USA.

4. *The ampere*

(i) *Ayrton–Jones balance*

The definition of the ampere, page 7, is simple and is useful for legal purposes, but if we tried to apply it directly to the realization of the ampere we would find that the infinite length of straight conductor, the negligible cross-section, even the inconveniently small force, cause insuperable difficulties. As however, at any rate for the purposes of this book, we accept the electromagnetic theory epitomized by Maxwell's equations, page 8, on which the definition of the ampere is based, we are entitled to select from that theory some formula suggesting a more practical system for realizing the unit.

A suitable starting point is Neumann's formula for the mutual energy W of two circuits s and s' carrying currents I and I'. It is

$$W = I I' M \qquad (3.20)$$

with M, called the mutual inductance, given by

$$M = 10^{-7} \int_s \int_{s'} \frac{ds . ds'}{r} \qquad (3.21)$$

where ds, ds' are vector elements of the circuits a distance r apart, the dot indicates the scalar product, and the integrals are taken round the circuits.

Energy is not a convenient quantity to measure, but force is. The force between the circuits in any direction x is

$$F_x = \frac{dW}{dx} = I I' \frac{dM}{dx}. \qquad (3.22)$$

There are some configurations for which it is possible to calculate

31

Fig. 3.16. Current balance of the National Physical Laboratory. One large coil has been lowered, so that the suspended coil can be seen. The weighing masses, platinum discs of about 4 g, are placed on or lifted off the scale pans by rods controlled by the knobs outside the case. (Crown copyright reserved.)

dM/dx, and which provide adequate force: one of them is a system of two coaxial helices, as in the Ayrton–Jones balance of the NPL, fig. 3.16.

Marble cylinders, cut with helical grooves to accommodate single-layer helices of bare copper wire, hang from the end knives of the beam. These cylinders are surrounded by larger ones supported on levelling screws and adjustable for height and horizontal position, and wound in two identical halves. When a suspended coil is coaxial with and half way between the two windings of the coil surrounding it, there is no force on it if the current flows in the same direction in the two windings of the large coil, but there is an axial force if the direction of the current in one of the windings is reversed. If, besides, the currents circulate in such a way that one suspended coil is urged up and the other down, and if equilibrium is restored by weighing masses, the weights give a measure of the current, provided the force per unit current on each coil can be calculated. It turns out that there is a convenient formula for the purpose, but its application entails a knowledge of the dimensions of the windings. The construction must therefore be such that the dimensions can be measured when desired, and this condition limits the windings to

32

single layers of bare wire.   The diameter which enters the calculation is that of the centre of the cross-section of the helix, obtained by subtracting from the outside diameter the average diameter of the wire, measured at the time of winding.   Not only the diameter, but also the axial position of each turn at several points round the cylinder, are measured, and the departures from nominal diameter and axial position serve to evaluate corrections to the force calculated from nominal dimensions.

A suspended coil is urged up or down not only by the current in the coil surrounding it but also, to a much smaller extent, by the other large coil.   The latter force however need not be known, it can be eliminated by taking a second weighing after reversing the current in all the windings at one end of the beam, for the ' direct ' force DI^2 is thereby unchanged, whereas the ' secondary ' force SI^2 is reversed.   In the NPL balance S is approximately 1% of D.

There are scale pans for weighing masses at both ends of the beam, and all errors that might be caused by gradients of fields of neighbouring magnets or currents are eliminated by reversing the currents in all the windings of both large cylinders.   The torque on the beam is thereby reversed, and the sum of the two masses, plus the small allowance for the change of rest point on reversal if the masses are not exactly right, corresponds to $2(D+S)I^2$ or to $2(D-S)I^2$ according to the positions of the switches.   If the sum of the four masses, corrected for rest points and buoyancy, is M, and the acceleration due to gravity is g, addition gives

$$4DI^2 = Mg \qquad\qquad (3.23)$$

from which I is calculated.

Although the effect of external magnetic fields is cancelled by reversal of the current in the large coils, that of magnetic material in the vicinity is not, for calculation of forces is in general possible only if the coils lie in a region where the magnetic susceptibility is zero over a volume much larger than their own.   All material intended not only for the coil formers but also for the balance and its pedestal must be tested and rejected if the susceptibility exceeds say 10^{-4}, or even 10^{-5} for components that are large or close to the coils.   (Magnetic susceptibility is defined as $\mu - 1$ where μ is relative permeability.)

Most parts of the NPL balance, including the heavy pedestal on which it rests, are of bronze, and the coil formers of marble.   Fused silica would be a better material still for the formers, because its thermal coefficient of expansion is only about a tenth of that of marble.

Each winding consists of two parallel helices starting at opposite ends of a diameter, the pitch of each helix being thus twice the pitch of the winding.   This arrangement not only makes possible a measurement of insulation between adjacent turns, but also provides the means of rendering one system of coils inoperative, whilst still maintaining current in it to equalize production of heat and convection of air on the two sides of the central pillar of the balance. For this purpose the current is sent in opposite directions in adjacent helices of one of the large cylinders.

Flexible connectors are required between the suspended coils, which move with the beam, and the rest of the circuit, which is fixed. There are four of them, two on each side, each consisting of 80 annealed silver wires 0·025 mm in diameter. Two of them are just visible in fig. 3.16, the other two are hidden by the nearly horizontal part of the arrester structure. The sag of the connectors can be adjusted for negligible effect on period and sensitivity.

Apart from the reduction of unsteadiness of rest point caused by dissipation of about 40 W in the balance case with the working current of just over 1 A, the symmetry of the double system of coils has the advantage that once the axes have been set vertical by means of spirit levels, the large coil can be adjusted for height and for coaxiality with the suspended coil, by observing the change of rest point as its position is varied. This adjustment can be performed with ease and without any necessity for keeping the current very exactly constant, only because it is possible to oppose the right hand and left hand systems to make the resultant torque almost zero.

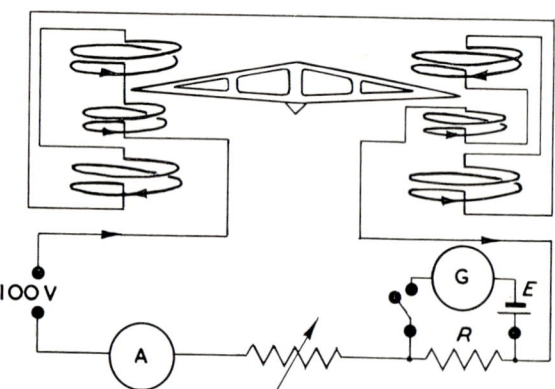

Fig. 3.17. Circuit for measurement, by means of a current balance, of E/R in amperes, where E is the e.m.f. of the Weston cell and R the resistance of the resistor.

The balance measures current to a few parts in 10^6, and as no pointer ammeter is capable of this precision, a potentiometer method is employed, as in fig. 3.17, to control the current under measurement. The potential drop across the resistor R is balanced against the e.m.f. of the Weston cell E by manual or automatic variation of the ballast resistance. Thus the current is maintained at a nominal value E/R, and what the weighings do is to determine the exact value of E/R in amperes for the cell and resistor employed, usually a single cell and a resistor of nominal value $1\ \Omega$.

(ii) *Rayleigh current balance*

The success of a balance of the Ayrton–Jones type depends on the possibility of accurate measurement of diameter and position at a large number of points on the coils. These measurements, which must be made to a fraction of a micrometre, can be very laborious: in the NPL balance, for instance, there is a total of 744 turns, and on one occasion nearly 1500 measurements of diameter and 4500 of axial position were made. Lord Rayleigh therefore, presumably with this drawback in mind, proposed a method which dispenses with linear measurements except for application of corrections. Suppose current flows in the same direction in coaxial circles, fig. 3.18, of radii r_1, r_2 an axial distance z

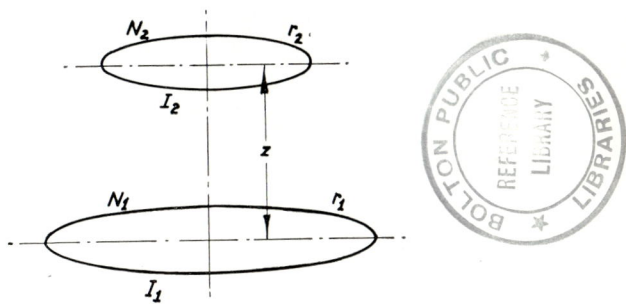

Fig. 3.18. Principle of Rayleigh-type current balance.

apart: the force between them can only be a function of r_1, r_2 and z. Also since it vanishes when z is zero, which is a position of equilibrium, and when z is infinite, there must be a value z_m in between for which it is greatest. But since that particular value z_m of z can only be a function of r_1 and r_2, the maximum force F_m, being function of r_1, r_2, z_m, reduces to a function of r_1, r_2. Moreover, as we can see from (3.21) and (3.22), dM/dx is dimensionless in length, thus F_m must be a function of the ratio of the radii, or

$$F_m = f(r_1/r_2). \qquad (3.24)$$

Rayleigh pointed out that r_1/r_2 can be obtained by a purely electrical experiment: if the circles, still coaxial, are placed in one and the same vertical plane in the magnetic meridian, currents I_1, I_2 in the circles produce flux densities proportional to I_1/r_1 and I_2/r_2 at the centre; if one of the currents is adjusted for a small magnet pivoted at the centre to remain undeflected on simultaneous reversal of both currents, these expressions are equal, and

$$\frac{r_1}{r_2} = \frac{I_1}{I_2}. \qquad (3.25)$$

As the ratio of two currents can be measured, for instance by a potentiometer, page 70, without any knowledge of the unit of current, we obtain r_1/r_2 without having made linear measurements.

In practice the force between single circles is too small, so it is increased by using closely wound coils of many turns. It is usual also to have two large coils, one above, the other below the suspended coil, because not only is the force thereby doubled, but the optimum distance between the two large coils, as well as their axial position relative to the suspended coil, can be adjusted by observations of the rest point when the forces oppose instead of adding, for the resultant force is then quite small and it is not necessary to keep the current constant to a high accuracy.

Formulae (3.24) and (3.25) are not exact when the windings have finite cross-sections, and as the correction to the force is not equal to the correction to the ratio of the radii, each must be calculated. But it is then found that if the ampere is to be realized to 1 or 2 in 10^6, calculation of the corrections to the required accuracy entails a fairly exact knowledge of the position of each turn of each coil, information which can only be obtained at the time of winding, and even then not very readily. Thus the proposal, although extremely elegant in principle, soon appears less attractive when applied to a practical system; it is interesting to note that a national laboratory which had a Rayleigh type balance is now using single-layer helical coils somewhat like those of the Ayrton–Jones balance.

Further reading:

Vigoureux, P. 1964. 'A determination of the ampere'. *Metrologia*, **1**, 3–7.

CHAPTER 4

realization of derived units

1. *The volt*

THE volt being defined, p. 8, in terms of the ampere and of the watt, its realization, to be effected according to the definition, would involve measurement of electric current and of power. But power and energy cannot be measured very conveniently or accurately. In this case a calorimeter method would presumably be used, but entails a knowledge of specific heats which again are not known to the precision with which electrical measurements can be made. It has therefore been the practice of national standardizing laboratories to select from electromagnetic theory some formula suggesting a more convenient and more accurate method of realizing the derived units, a course they are fully justified in adopting, since the definitions of the derived electric and magnetic units are consistent with, and indeed based on, that theory.

In the case of the volt, for instance, one expression which comes to mind is that for the force between two metallic parallel plates maintained at different electric potentials. If the capacitance between the plates is C, the potential difference V and the coordinate perpendicular to the plates z, the energy is

$$W = \tfrac{1}{2}CV^2 \qquad\qquad (4.1)$$

and the force in the direction z is

$$F_z = \frac{\mathrm{d}W}{\mathrm{d}z} = \tfrac{1}{2}V^2\frac{\mathrm{d}C}{\mathrm{d}z}. \qquad\qquad (4.2)$$

There are many configurations for which $\mathrm{d}C/\mathrm{d}z$ can be calculated. For two parallel plates of equal area A a distance a apart the capacitance, if edge effects are eliminated by guard rings, is $\epsilon_0 A/a$ and the force is

$$F = -\frac{V^2\epsilon_0 A}{2a^2}. \qquad\qquad (4.3)$$

The electric constant ϵ_0 is known from (2.3) in terms of the speed of light, the force can be measured if the upper plate is suspended from one of the end knives of a balance, and the distance a, which must be kept small to secure adequate force and to minimize edge effects, can be determined by interferometry. The National Standards Laboratory of Australia (NSL), ever in the forefront of development in electric measurements and standards, is trying the ingenious idea of using mercury for the lower ' plate ', and observing the displacement of the surface against gravity when the potential difference is applied.

Whilst direct realization of the volt is desirable as a check, it is not essential because the volt can be determined from the ohm and the ampere by a purely electrical measurement of high precision, and it happens that at least two good methods of realizing the ohm are available. It is to these that we now turn.

2. *The ohm*

Since the reactance of a self or mutual inductance, ωL or ωM, and that of a capacitance, $1/\omega C$, have the dimensions of resistance, as can be seen from the definitions of the henry and the farad, page 8, and their dimensions, page 10, they can in principle be compared with resistance in some network or bridge circuit. If moreover L or M or C can be calculated from their linear dimensions, the ohm could be realized, since the frequency involved in ω is easily measurable to high accuracy. It is true that the result would be the resistance to alternating current, which because of skin effect may differ from what it is for direct current, but it is possible to use a comparatively low frequency, say 100 Hz or less, it is possible to make resistors which vary very little in resistance with frequency and to calculate the variation, and alternatively, with suitable provision for repeating the bridge balance at a few frequencies, it is possible to determine the correction required, if any, to render the result applicable to direct current.

3. *The calculable mutual inductor*

Self and mutual inductors have both been employed to realize the ohm but it is probably true to say that the effect of the leads is more difficult to allow for in the case of the self inductance. Here we confine ourselves to describing the mutual inductor of the NPL and to explaining how it serves to realize the ohm.

Two identical coils, fig. 4.1, separated by a distance approximately equal to twice the length of each, make up the primary; the secondary is a coil of larger diameter in a plane midway between them. Accurate measurement of diameter and axial position can be made only on single-layer windings. The primary is therefore wound in a single layer with bare wire in helical grooves cut in the cylindrical former of fused quartz, a stable material of small thermal coefficient of expansion, about 0.5×10^{-6}/K.

When both primary windings carry current in the same direction the lines of force bulge out in the region between them. Their general shape is as for the Lorenz apparatus, fig. 4.8. Albert Campbell, who designed the first of these inductors at NPL, pointed out that the most favourable secondary coil was one whose central filament just included

Fig. 4.1. Calculable mutual inductor of the National Physical Laboratory. The primary coil is wound in a single layer in a helical groove on a former of fused quartz. The secondary coil lies in a channel of square section in a glass disc. (Crown copyright reserved.)

all the lines linking the primary but none of the return lines. The diameter, calculated from the dimensions of the primary, is an optimum not only because the mutual inductance is then the maximum obtainable from a given primary, but more important still, because the coil then lies in the toroidal region of nearly zero field. This property is extremely useful for, in order that the mutual inductance may be large enough to have a reactance easy to compare with resistances in a network, say about 10 mH, the secondary must be wound with four or five hundred turns which makes an estimate of diameter less accurate than for the primary, and entails application of a correction for the finite size of the cross-section of the winding. But if the central filament has the optimum diameter, the correction is least, and its uncertainty is least also.

The secondary coil rests on 3 levelling screws and the position of its centre can be changed laterally by 3 other screws. Coincidence of its

39

axis with that of the primary is secured by mechanical means, and position along the common axis by passing current in opposite directions in the two parts of the primary and adjusting the levelling screws for zero mutual inductance.

Neumann's formula (3.21) for the mutual inductance leads to a fairly simple expression when one of the circuits is a helix and the other a coaxial circle, in this case of diameter equal to that of the central filament of the secondary. Convenient formulae are also available to correct for the finite cross-section of the secondary, and for departures of the diameter and of the axial position of each turn of the primary from the nominal values.

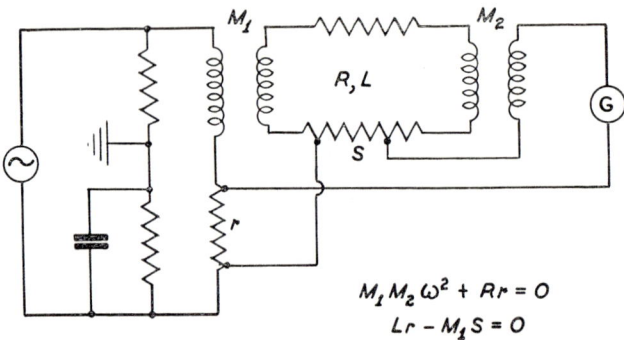

$$M_1 M_2 \omega^2 + Rr = 0$$
$$Lr - M_1 S = 0$$

Fig. 4.2. Campbell's network for realizing the ohm in terms of length and time.

The network to realize the ohm, fig. 4.2, also devised by Albert Campbell, requires in addition to the standard, a variable mutual inductor, also Campbell's design, previously calibrated against the standard. The main and auxiliary conditions of balance are effected by alternate adjustments of M_2 and S, and since the two mutual inductances are known and the frequency can be measured very accurately, the product Rr is obtained in terms of the metre and the second, both defined by atomic constants as explained in the previous chapter. Since R, the resistance of the loop formed by the two primaries and the auxiliary resistors, and r are fixed resistances each of which can be compared with the laboratory standard of resistance, the value of the latter is obtained in ohms. The probable error of this realization is 2 in 10^6, due mostly to uncertainty of the linear dimensions of the standard inductor.

4. The calculable capacitor

Until 1956 nobody knew how to make a capacitor whose capacitance could be calculated with the accuracy desired for realizing the ohm. In that year Thompson and Lampard of NSL, Australia, showed how to do so in principle, they and their colleagues soon after made the instrument, and they have realized the ohm by its means to an accuracy never before attained.

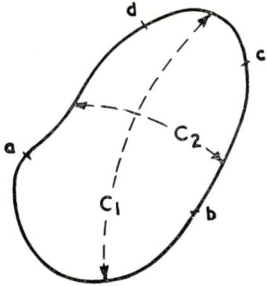

4.3.   Lampard's theorem: the capacitance per unit length between opposite quadrants of the infinitely long cylinder is independent of the size or shape of the cross-section provided the segments are chosen to make the capacitances equal.

A theorem of Riemann on conformal transformations can be used to prove that if an infinitely long right cylinder of any cross-section is divided into four segments, as in fig. 4.3, and the internal capacitances between opposite segments per unit length of the cylinder are denoted by C_1 and C_2,

$$\exp\left(-C_1\pi/\epsilon\right)+\exp\left(-C_2\pi/\epsilon\right)=1 \tag{4.4}$$

where ϵ is the permittivity of the medium enclosed by the cylinder.   If the segments are chosen so as to make C_1 equal to C_2, the capacitance per unit length becomes

$$C=\frac{\epsilon\log_e 2}{\pi} \tag{4.5}$$

independently of the shape and size of the segments.
    If the interior of the cylinder is a vacuum, ϵ is the electric constant ϵ_0, and from (2.3) and (4.5)

$$C=\frac{\epsilon_0\log 2}{\pi}=\frac{10^7\log 2}{4\pi^2 c^2}. \tag{4.6}$$

If in (4.4) we write C_m for $(C_1+C_2)/2$ and α for $(C_1-C_2)/(C_1+C_2)$ we find by expanding in series in powers of α

$$C_m=C\left(1+\frac{\alpha^2}{2}\log 2\right) \tag{4.7}$$

which shows that the error made by taking C for C_m is

$$\frac{C_m-C}{C}<\frac{(C_1-C_2)^2}{11C^2}. \tag{4.8}$$

In practice the capacitor, whose cross-section is shown in fig. 4.4, consists of four insulated metal cylinders A, B, C, D nearly touching each

41

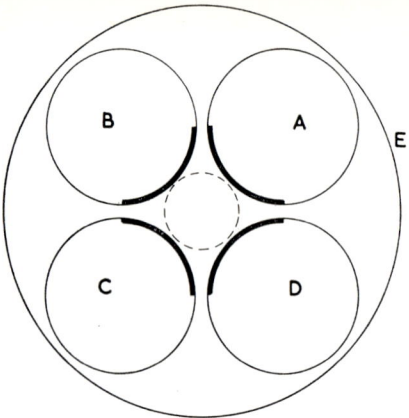

Fig. 4.4.   Cross-section of calculable capacitor of Thompson and Lampard.

other, surrounded by an earthed screen E also close to the cylinders. This design provides gaps of re-entrant shape; thus if one cylinder is raised in potential and the other three are earthed, no line of force emerging from the outer segment of the first reaches the diagonally opposite cylinder.   In consequence the relevant capacitance between A and C is only that of the inner space between the heavily marked segments.   The system cannot however be infinitely long, and when it is finite the length by which to multiply the capacitance per unit length in (4.6) is uncertain because of end effects.   Accordingly two metal tubes, both earthed and just clearing the inside segments of the first four cylinders, are inserted one from each end; their trace is dotted in the figure.   They screen opposite cylinders, say A and C, one from the other, except in the interval separating them.   The ends facing each other are terminated by flat polished plates which form the partial reflectors of a Fabry–Pérot optical interferometer.   Even this interval cannot be taken as the length of the capacitor because the electric field is distorted near the end faces, and the lines of force there do not lie wholly in planes perpendicular to the axis, as assumed by (4.4).   But if, after the capacitance for some interval is compared with the laboratory standard, one cylinder is moved inwards a distance l, determined by a count of fringes as described on page 17, and a second comparison made with the laboratory standard, the end effects are eliminated and a calibration of the interval observed on the laboratory standard is obtained, since the difference is known to be l times C of (4.6).

The formula is exact only if the cylinders are made and assembled so well as to make C_1 equal to C_2, but in practice perfect symmetry is difficult to attain.   If however the cylinders are adjusted to reduce to a negligible value the error given by (4.8), the mean of C_1 and C_2 may be taken to correspond to C in (4.6).

42

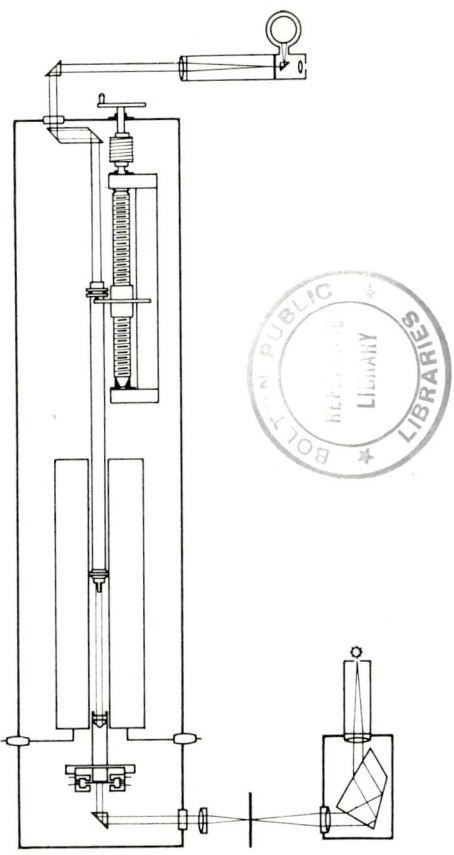

Fig. 4.5. Optical system of the calculable capacitor of the National Standards Laboratory.

The optical system employed at NSL for measuring the displacement of one of the inner screening tubes is shown in fig. 4.5. The collimated light enters the vacuum chamber by the window at the bottom, is deflected vertically upwards by the prism, enters the lower central tube, goes through the semi-transparent film of the lower interferometer plate, and after multiple reflections between the lower and upper faces passes inside the upper central tube. It emerges at the other end, passes through prisms and a window at the top of the vacuum vessel and enters the telescope. In the image plane of the telescope objective is a pair of cross-wires and, immediately after, a diaphragm with two apertures, one large for viewing the full field, the other small to exclude everything but the centre of the fringe system. When the small aperture is in position a prism deflects the emerging light into the photomultiplier. As an

43

integral order of interference is all that is required in this case, it is sufficient to adjust the position of the upper tube, by the lead screw and handle of fig. 4.5, to maximum intensity of the central fringe. To perform this adjustment by the photoelectric method the lower tube, mounted on a disc spring, fig. 4.5, is vibrated at very small amplitude by an electromagnet at 30 Hz, and the output of the photomultiplier is fed to an amplifier tuned to that frequency, followed by a cathode-ray oscilloscope whose horizontal sweep is operated by the same 30 Hz supply. At the correct setting of the movable tube the output of the photomultiplier contains no fundamental but only d.c. and even harmonics, which the amplifier tuned to 30 Hz does not transmit, thus the CRO gives a null reading.

The change of capacitance calculated from (4.6) is about 0·4 pF for a 20 cm displacement of the upper cylinder, and although a laser would yield good fringes for much larger separations, the overall length of the capacitor is limited by convenience and by the required rigidity of its electrodes. The wavelength of a helium-neon laser is 620 nm and the number of fringes, $2l/\lambda$, about 6×10^5, so that a measurement of length to 1 in 10^8 entails a precision for each of the two settings of about 1/400 of a fringe, which the photoelectric detector seems to provide.

The small capacitance is a drawback because the convenient value for comparison with resistors in a bridge is of the order of 1 to 10 nF. It is therefore necessary to 'build-up' to the desired large value in four 10 to 1 steps by means of capacitors of good short-term stability. Transformers of known ratio developed at NSL, NPL and elsewhere are well suited to this purpose. Figure 4.6 shows the 10 : 1 tapped secondary

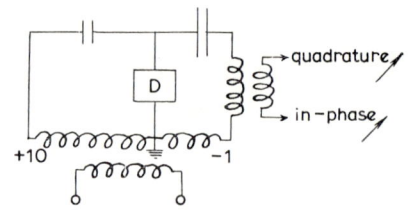

Fig. 4.6.   10 : 1 ratio transformer bridge with small adjustable in-phase and quadrature voltages for balancing.

and the auxiliary transformer whose primary is fed with small voltages, also derived from the main primary via 2 additional secondaries and an RC network, to provide in-phase and quadrature adjustments for balance.

When a capacitor C_1 of 5 nF has thus been standardized in terms of the calculable capacitor, a resistor R_1 must be measured in terms of it and frequency, and compared with the 1 Ω standards which serve to maintain the unit in the laboratory. For convenience an angular frequency of 10 000 rad/s is chosen, to make the reactance $1/\omega C_1$ equal to 20 kΩ. The

44

first operation is then to measure a resistance of 20 kΩ in terms of that reactance. If the resistor and capacitor are connected in series and voltages V_1, V_2 approximately equal, in quadrature, and having a common point, are applied to the ends of the series impedance, the junction of R_1 and C_1 can be brought to the potential of the common point by a small adjustment of V_1, and since the currents in the two components are then equal

$$\frac{V_1}{R_1} = V_2 \omega C_1. \qquad (4.9)$$

In 1912 Albert Campbell had balanced a resistance against the reactance of a mutual inductor by this method, using as supplies the two voltages of a 2-phase alternator. It is not however easy to produce, by this or other methods, voltages accurately known and in quadrature, but

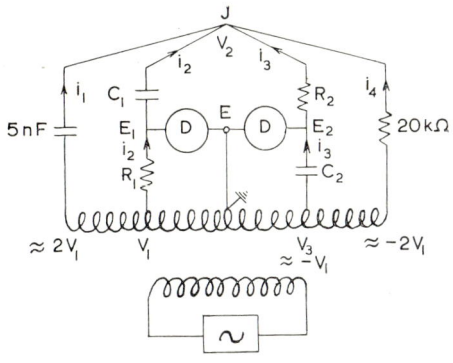

Fig. 4.7. Transformer bridge for measurement of resistance in terms of capacitance and frequency. D, D are null detectors.

if the system is duplicated, as in fig. 4.7, by another known voltage V_3 and components R_2, C_2, we have

$$V_3 \omega C_2 = \frac{V_2}{R_2} \qquad (4.10)$$

giving by division

$$\omega^2 C_1 C_2 R_1 R_2 = \frac{V_1}{V_3}. \qquad (4.11)$$

The quadrature voltage V_2 can be produced from the bridge supply and an RC network, but it need not be very accurately known as it does not enter relation (4.11), at least to the first order. In fig. 4.7 the

45

transformer supplies, in addition to V_1, V_3, two equal and opposite voltages $\pm 2V_1$, which feed a resistance and a capacitance equal to those being compared. If by small adjustments of these two components and of V_3 the junctions E_1, E_2 are brought to Earth potential, the voltage V_2 of the junction J will have the desired value, as can be seen by writing that the sum of the four currents entering the junction J is equal to zero. For simplicity, since V_1 and V_3 are nearly equal and opposite, and since the impedances of all the components are equal except for phase, we may write x for V_2/V_1 and 1 for V_1, R and $C\omega$. The currents, starting with i_1, are $(2-x)$j, 1, $-$j and $-2-x$, whose sum is zero if x is equal to j, i.e. if V_2 is equal to jV_1.

As the components are all close to nominal, only small in-phase and quadrature voltages, produced as in fig. 4.6 by an auxiliary transformer, need to be added to V_3, i.e. to $-V_1$, to secure the balances, which yield (4.11), apart from corrections due to phase defects.

The resistors R_1, R_2 in parallel are next compared, also by a transformer bridge, with a specially made 10 kΩ resistor whose change with frequency can be calculated or measured, and whose resistance is compared with the 1 Ω d.c. laboratory standard via a build-up process, for example that of Hamon, page 69.

This method of realizing the ohm thus involves several consecutive operations: measurement by interferometry of the displacement of the screening cylinder of the calculable capacitor, measurement of a fixed capacitor of the same small value, say about 0·5 pF, in terms of the calculated capacitance, four measurements on a 10 : 1 ratio bridge to determine the capacitance of two 5 nF capacitors, measurement on a capacitance-resistance bridge of two resistors of 20 kΩ in terms of the 5 nF capacitances and of frequency, comparison of the two resistors in parallel with a 10 kΩ resistor suitable for a.c. and d.c., and measurement of the 1 Ω laboratory standard in terms of the 10 kΩ resistor by a ' build-up ' process. It is surprising and remarkable that the combined uncertainty of so many measurements should be assessed at no more than 2 in 10^7, and when several laboratories will have made calculable capacitors and the associated equipment, it will be interesting to see to what extent their results substantiate this claim.

All realizations of the ohm by this method are subject to another probable uncertainty of some 5 in 10^7 because the square of the speed of light enters formula (4.6) for the capacitance. This uncertainty does not however affect comparison of results provided the same value of c is used by all, and it will decrease should measurements of the speed of light, now being made, improve the accuracy with which the constant is known.

It seems possible to maintain the capacitor and the associated apparatus in good adjustment more or less permanently, and to carry out all the operations described above in one day, whereas some months are needed to measure the linear dimensions of inductors. The equipment would therefore appear to be as well adapted for maintenance as it is for realization of the ohm.

Further reading:
Clothier, W. K. 1965. ' A calculable standard of capacitance '. *Metrologia,*
  1, 35–56.
Thompson, A. M. 1968. ' An absolute determination of resistance based on a
  calculable standard of capacitance '. *Metrologia,* **4,** 1–7.

5. The method of Lorenz

In 1873 L. V. Lorenz proposed a method of realizing the ohm which, although not capable for the reasons given below of an accuracy comparable with that provided by the calculable capacitor or even the calculable inductor, is nevertheless mentioned because of its historical and more especially of its tutorial interest.

An apparatus of the Lorenz type, incorporating several novel features, was built at the NPL by F. E. Smith, who used it in 1913 to realize the ohm. A long shaft coupled to an electric motor carries two discs; on each side of each disc there is a stationary coil, fig. 4.8, coaxial with the

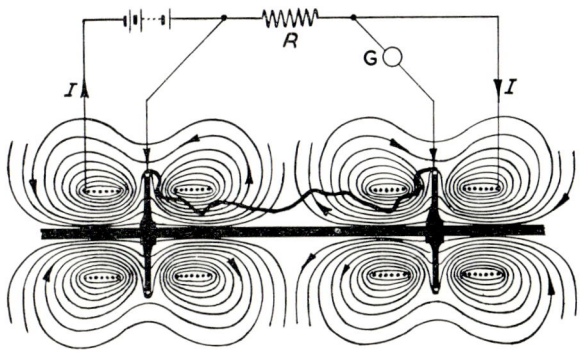

Fig. 4.8. Principle of Lorenz' method of measuring resistance in terms of
length and time.

shaft and wound with wire resting in helical grooves in the marble former. The construction, similar to that of the coils of the current balance, page 32, allows accurate measurement of the dimensions. If in addition the diameters of the rims of the discs are known, and if the distance from coil to coil is measured before and after each experiment, the mutual inductance M of the coils and the rims of the discs can be calculated. The rims are insulated from the discs and shaft and are joined by a conductor, which in the figure is shown by an irregular line to emphasize that its actual position is immaterial provided it threads the two inner coils. In practice it is led down a groove in the disc, along a hole in the shaft and up a groove to the rim of the other disc. If the shaft is rotated exactly one turn, the conductor cuts all the flux linking one disc or the other, i.e. MI, where I is the current in the coils; if the shaft is spun at a speed n the e.m.f. induced in the conductor is, by Faraday's law of electromagnetic induction, MIn.

Current is provided by a battery of large accumulators and the circuit is completed by the resistor R in series with the coils. The potential terminals of R are connected to brushes at the rims of the discs in such a sense that the p.d. across R opposes the e.m.f. generated in the conductor. If the current in the coils and in the resistor is reversed, MIn and RI are both reversed, and so is the galvanometer deflection, which indicates their difference. Thus if the resistor is shunted or the speed adjusted for the galvanometer reading to remain unchanged on reversal of the current, RI is equal to MIn, i.e. R is equal to Mn, and is thus obtained from the linear dimensions of the system and the rotational speed, measured by a chronograph at the far end of the shaft, in other words in terms of the metre and the second. The shaft is about 9 m long because the motor, which contains iron, must be sufficiently far away for its effect on the mutual inductance to be negligible.

Duplication of discs and coils has many advantages: the e.m.f.s induced by the Earth's magnetic field tend to cancel; the non-random parts of the contact and thermal e.m.f.s at the brushes are in opposition; the rather small induced e.m.f., 2 mV in each system, is doubled. Another interesting feature is that, as for the mutual inductor, page 38, the proportions of the system are chosen such that the rim of the disc lies in zero magnetic field; a small error in the measurement of the diameter of the rim thus causes nothing like a proportional error in the final result.

The beauty of the method lies in its simplicity in principle and in its being the only continuous current method. On the other hand the random errors, and more so still the errors in measurement of the inter-coil distances, give rise to a probable uncertainty of 5 in 10^6.

Further reading:

Smith, F. E. 1914. 'Absolute measurements of a resistance by a method based on that of Lorenz'. *Phil. Trans A*, **214**, 27–108.

1. *Material standards*

REALIZATION of a unit presupposes an instrument capable of preserving the result. The base unit, the ampere, might be preserved by an ammeter in series with the coils of the current balance. As however pointer instruments can normally be read only to 1 in 1000, and as even reflecting instruments, which employ an optical system instead of a pointer, are at best reliable to no more than 1 in 10^4 whereas the probable uncertainty associated with the current balance is as low as 4 in 10^6, reflecting and pointer instruments alike would be of little help for maintenance of the unit.

Likewise if the volt were realized by means of an electrometer to one part or two in 10^6, some device other than a pointer or reflecting voltmeter would be needed to preserve it.

Scientists have been fortunate in having had at their disposal for many years two instruments which overcome the difficulty: they are the standard resistor and the standard cell. We have seen, fig. 3.17, how the resistor and cell, connected in the circuit of the current balance, not only serve to keep the current constant during the weighings, but are in effect calibrated by the balance, which determines the quotient E/R, so that these two material standards preserve the ampere realized by the balance.

The standard resistor by itself serves to maintain the ohm realized by any of the methods described in the preceding chapter, and should the electrometer prove successful at realizing the volt, the Weston cell, in conjunction with resistance dividers, could be used to preserve the unit in the same way that it now does, except that it would then be calibrated directly, whereas at present its e.m.f. is deduced from independent realizations of the ampere and of the ohm.

Standard resistors are made of annealed wire of an alloy of low temperature coefficient of resistance, low thermal e.m.f. against copper, and great stability, wound with as little constraint as possible over an insulating frame, or ' former ', fixed to the underside of a plate with insulating washers through which pass leads of stout copper wire or rod. The resistance wire is soldered to these leads, whose other ends are provided with terminals. In addition to these ' current ' terminals ' potential ' terminals are connected to points on the resistance wire or on the current leads. The method of measurement yields the resistance between the junctions of these potential leads with the resistance wire or

49

Fig. 5.1. Standard resistors with current and potential leads. (Crown copyright reserved.)

current leads, a resistance not affected by possible variations of contact resistance at the terminals. Two types of these 'four-terminal resistors' are shown in fig. 5.1. The resistance wire is often made of the alloy manganin, containing approximately 85% copper, 11% manganese and 4% nickel, and the former on which it is wound is immersed in oil or in a dry gas in a can closed by the insulating plate.

If the many precautions indicated by experience are taken both in the preparation and heat treatment of the manganin used for the element, and in the construction, maintenance and use of the resistors, it is possible to obtain a stability of 1 in 10^7 per year, or better, after a few years' ageing of the resistors.

For work of the highest precision it is impracticable to use a single resistor as the reference standard, because no single resistor can be presumed to have the necessary constancy of resistance over a long period of years. Constancy of value from year to year can be judged by the relative constancy of several resistors, but it cannot be established with certainty. As all known resistors drift slightly relative to one another from year to year, the reference standard of national laboratories consists of a group of a few resistors, perhaps five or more, of nominal value 1 ohm, that have shown the greatest relative constancy during the preceding years. The average value for this group is assumed to have remained constant during the period, and this assumption, together with the relative values at any particular time, determines the individual values at that time. The average value is determined by an absolute measurement, but this measurement is subject to an uncertainty of perhaps 1 part per million. For international comparisons a precision of 1 in 10^7 is both desirable and practicable, for reference standards from two countries can be compared with at least this precision. It is customary therefore to assign values to the reference standards with a precision of 1 in 10^7, and to obtain international uniformity to this precision or as near to it as practicable, while recognizing that the unit thereby adopted may differ from the true unit by as much as a part per million.

New resistors are usually less stable when first made than after a few years, and old ones tend to develop instability with age. Resistors of various ages are examined when selecting the reference group, and the behaviour, considered in relation to their known structure, serves as a guide to the construction of new standards.

The Weston cell is a primary battery in which the positive electrode is mercury, the negative electrode an amalgam of cadmium and mercury, and the electrolyte is cadmium sulphate. Mercurous sulphate paste over the positive electrode serves as depolarizer. In the ' saturated ' cell, fig. 5.2, which is the type normally used to maintain the volt, cadmium sulphate crystals keep the solution saturated at all working temperatures, and it is usual for the electrolyte to be made slightly acid, for instance 1/40 molar.

Saturated Weston cells, if made to the same specification as regards procedure of assembly and quality of the component materials, are equal in e.m.f. to a few parts in a million, in a group made on one occasion, and to 1 or 2 parts in 10^5 for groups made at different times, but can be stable to a part in a million per year, or better.

They are compared with one another to a precision of better than 1 in 10^7, but no single cell can be relied upon to remain constant to this degree over a period of years. The reference standard of a national laboratory would therefore consist of a large group, say twenty cells or more, made at different times during the preceding thirty years or so. The cells forming the group are selected as those showing the closest relative constancy in comparative measurements; the average value for the group is assumed to have remained constant since the last time it was determined by an absolute measurement. Values based on the absolute measurements are assigned to 1 part per million, although, just as with standard resistors, it is recognized that the unit thereby adopted may differ from the true unit by several parts per million.

Batches of cells are made or bought at intervals of two or three years. New cells appear to stabilize during their first year or two, but are liable to gradual deterioration. Old cells are removed from the reference group as soon as signs of deterioration appear, and each is replaced by a fairly new cell that has reached the stable stage. Due allowance is of course made for any difference between the old and new cell when assigning values to the cells on the assumption of a constant average; the assumption is merely of constancy with time, not constancy irrespective of change in the identity of members of the group.

The standard resistor and the Weston cell enable us to measure not only resistance and potential difference by comparison, but also electric current and electric power by comparatively simple methods. The cell is however an exclusively direct-current device, and the standard resistor described above is not suitable for alternating current, at any rate at high frequencies. The cell and resistor do however serve to calibrate instruments which provide electric standards for use with alternating current.

The Weston cell is even now the material standard almost exclusively used for day-to-day maintenance of the volt and for international comparisons, and cells are available enclosed in containers which keep their temperature constant to close on 0·001 K even during surface or air travel. A recent development might nevertheless provide a more robust standard of potential difference, which would eliminate transport of delicate cells for comparison between laboratories.

This instrument is the voltage-regulator or Zener diode, a silicon semi-conductor device, the potential drop across which is almost independent of the reverse current. In its simplest possible form as a reference standard the diode would be fed from a battery, in the reverse direction, through a high resistance, and the potential drop across its terminals,

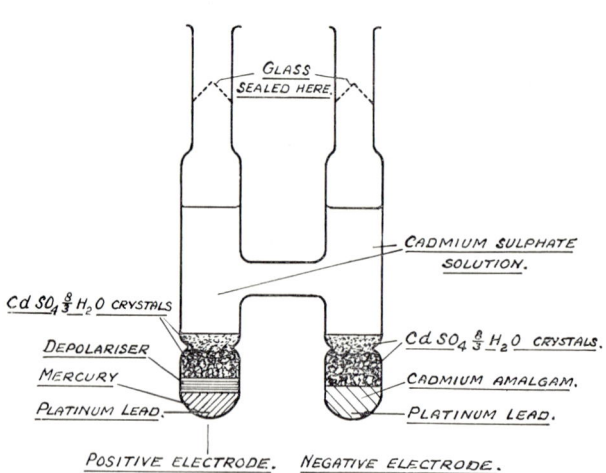

Fig. 5.2.   Saturated Weston cell.

52

determined by means of a Weston cell and potentiometer, would provide a reference voltage. For best results the circuit would be more complicated, incorporating arrangements to ensure an almost invariable diode current, for the potential drop does depend on the current, although to a very small extent only.

Voltage regulator diodes are available in a great range of voltages, from 4 to about 70, but those which have been suggested as standards of electric tension vary between 6 and 12 volts. In precise work the Zener diode has not yet proved an adequate substitute for the Weston cell.

In the preceding chapters we described calculable inductors and capacitors for realizing the henry or farad, and the ohm. Inductors and capacitors are also made into stable, although not calculable instruments, which laboratories have exchanged to compare their working units.

2. Drift of material standards

Although the best standard resistors and Weston cells drift only very slowly, it remains true that, after ten years or so, the values might have changed by 1 or 2 parts in a million. Some idea of the drift can be obtained from international comparisons, made every two or three years, of resistors and cells of national laboratories, but as experience of over half a century has indicated the best types of resistors and cells, the material standards of the various laboratories are now so much alike that it would be unsafe to assume that the comparatively small relative changes they exhibit are not slight irregularities of a much larger drift they might all have in common. From time to time resort must therefore be had to some independent check of their stability.

Until recently these checks were provided by absolute determinations of the ohm and the ampere. Measurements with a standard mutual inductor or some other calculable reactance would check the ohm possibly to 1 part in a million, and the current balance would give the ampere, or E/R, with a probable error of 4 parts in a million, and therefore E to about the same accuracy. Apart from the uncertainty, which is comparable with, if not even greater than, the drift of good material standards in several years, these absolute determinations occupy many months of the time of highly skilled persons, so that it has not been found practicable to undertake them at intervals shorter than ten years or so.

The calculable capacitor provides a quicker and simpler method of measuring resistance than is afforded by the mutual inductor, and is likely to prove as convenient for periodical checks of resistance standards as it is for realizing the ohm, but it remains true that a check of the volt via the current balance is an arduous and lengthy procedure. It is fortunate therefore that recent work on the determination of atomic constants comes to our aid in supplying more rapid and convenient checks of the ampere and the volt.

C

3. *Use of atomic constants for control*

Physicists are interested in accurate measurement of atomic constants not only for the sake of the self-consistency it tends to establish, but also because comparison of results obtained by different methods may lead to a better understanding of nature. It happens, for instance, that an important, dimensionless, constant which links c, e and h, the fine-structure constant, $\mu_0 e^2 c/2h$, can be determined by two independent methods, which give values differing by about 5 times the combined probable uncertainty of the two determinations. As one of these methods involves several fairly large theoretical corrections, the lack of agreement suggests that the theory on which the corrections are based should be critically scrutinized if not revised.

Apart from this general scientific interest, measurement of atomic constants finds an application in the maintenance of units preserved by material standards, for since these constants are thought, or assumed to be, of their nature unchanging, once one has been determined with high accuracy, the value can be adopted at least provisionally, and the method by which it was measured used in reverse to check the value of a resistor, a capacitor, a Weston cell or some other material standard.

We give two examples.

4. *The Josephson effect; the volt*

B. D. Josephson predicted in 1962 that two superconductors separated by a thin gap would exhibit rather unexpected phenomena, which were however promptly verified by experiment. For the application which concerns us the Josephson effect can be described as follows:

In fig. 5.3, A and B are superconductors separated by an insulating gap, which might be an oxide film say 1 to 2 nm thick. If A and B are connected to a battery and variable resistance, an electric current flows, unlike what would in general happen if they were ordinary conductors, because electrons in superconductors are capable of 'tunnelling' through the 'barrier'; in spite of the flow of current there is no potential drop across the gap, and the current is equal to E/R where E is the e.m.f.

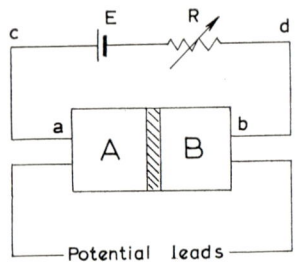

Fig. 5.3.   Principle of the Josephson effect.

54

of the battery and R the resistance of the portion *acdb* of the circuit. In fig. 5.4 the graph of p.d. against current would be the $X'OX$ axis. If however the junction is irradiated with electromagnetic waves, and the current is varied as before, there appears between A and B a potential difference which changes abruptly in steps V given by the equation

$$2eV = hf \qquad (5.1)$$

where f is the frequency of the electromagnetic radiation and e and h are the electronic charge and Planck's constant. The factor 2 in the equation occurs because in superconductors electrons act in pairs. The steps are represented in fig. 5.4 by the line ZOZ'.

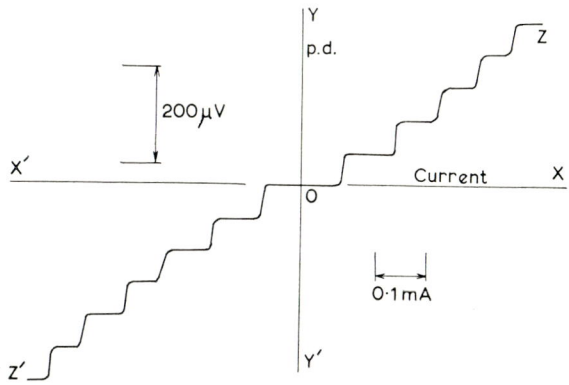

Fig. 5.4.   Voltage-current characteristic of a Josephson junction irradiated by electromagnetic waves of 11 mm.

One method of making a junction is to heat a niobium wire to produce a thin oxide film on its surface, then transfer a drop of solder to it. A copper wire is next embedded in the solder, and one copper and one niobium lead serve as current leads, the other pair as potential leads, as in fig. 5.5. The junction is placed inside and near the end of a wave guide immersed in liquid helium in a Dewar vessel, and excited by a klystron or other generator of microwave energy. A short-circuiting plunger at the end of the guide is adjusted for maximum microwave electric field at the junction.

Equation (5.1) not only offers a means of measuring the important atomic constant $2 e/h$, but also provides a method of verifying from year to year the stability of the material standard of electric tension of the laboratory, usually a Weston primary cell. This verification does not even require an exact knowledge of the value $2 e/h$, but only that the same value be used each time.

The beauty of this method of measuring the drift, if any, of the material standard is the simplicity of the frequency/voltage relation (5.1), which requires no corrections. On the other hand the measurement is not easy, because 2 e/h being $483 \cdot 5941 \times 10^{12}$ Hz/V (this figure refers to the volt as maintained at the NPL in 1970), the potential steps are small even for microwave frequencies, about 78 μV for an 8 mm wave, and comparison of this small potential difference with an e.m.f. just over 1 V is not easy to accomplish with a precision of 1 in 10^6, a precision all of which is needed because good Weston cells do not drift by more than that amount in a year or so.

Two ways of rendering the comparison less arduous suggest themselves: instead of using a single step it is possible to use several, indeed 500 have been used. Alternatively the junction can be irradiated by waves of the order of 1 mm length instead of 10 mm, which make the step nearer 1 mV, much easier to compare with 1 V to the desired accuracy.

Both methods however have limitations: if the number of steps is large there may be some difficulty in keeping the junction biased by the current on one and the same step during the measurement of voltage; on the other hand long-wave lasers, for instance the $0 \cdot 3$ mm hydrogen cyanide laser, are not easy to stabilize from a caesium beam or an equivalent frequency standard. Recently however the first method has been used with success on 500 steps of 20 μV each, to yield about 10 mV.

To compare a voltage of say 200 μV, obtained from a few steps, with the $1 \cdot 02$ V of the Weston cell, a potential divider is necessary. The divider may have a fixed total resistance R_c and an adjustable tap to yield a variable ratio of $0 \cdot 001$ or less. The divider current, provided by an accumulator, is first adjusted by a series variable resistance so that the p.d. across R_c balances the e.m.f. of the Weston cell; next the tap is adjusted for the volt drop across the tapped portion R_J to balance the known p.d. E_J across the junction. The e.m.f. of the cell is then $E_J R_c / R_J$.

As however a variable voltage divider is still more difficult to make than a satisfactory fixed divider, an alternative method is sometimes employed in which as before a current, I_c say, is adjusted by an external variable resistor to balance the e.m.f. E_c of the cell. The current is then altered, to I_J say, to balance the drop across the fixed low resistance portion R_J of the divider against the p.d. E_J provided by the junction. The e.m.f. of the cell is then $E_J R_c I_c / R_J I_J$, in which the fixed ratio R_c / R_J is known from the calibration of the divider, and the ratio I_c / I_J of the currents still has to be determined. Although a potentiometer, page 70, is normally used to compare potential differences, it can be applied to comparison of currents, in principle as follows: it is connected, fig. 5.5, in series with the divider and variable resistor, and after adjustment of I_c a balance is obtained against the auxiliary Weston cell E by means of the potentiometer tap or dials. Likewise after adjustment of I_J another potentiometer reading is obtained by balancing against E, which need not be known, but must not change during the experiment. The required

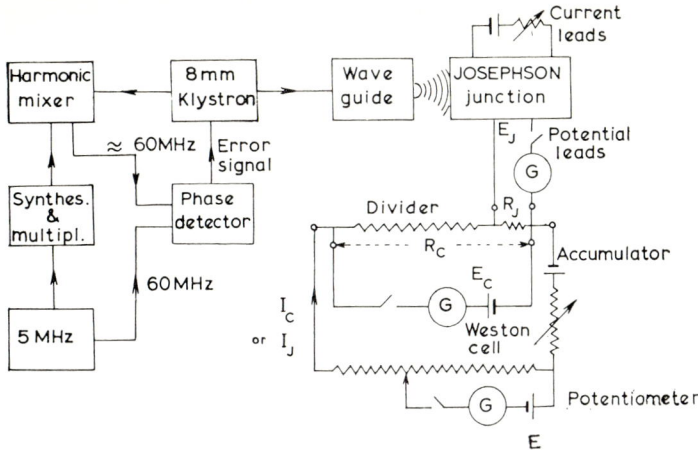

Fig. 5.5. Simplified diagram of equipment for irradiating Josephson junction with electromagnetic microwaves and for comparing the junction voltage $hf/2e$ with the e.m.f. E_c of a Weston cell.

ratio of currents is the inverse ratio of the dial readings. Even with a fixed divider the potentiometer can be dispensed with if E_J itself is adjusted for balance by varying the radio frequency.

Figure 5.5 is a simplified block diagram of the equipment. The klystron is a generator whose frequency can be altered to a slight extent by adjusting the d.c. voltage on its reflector guide. The synthesizer and associated multiplier provide a frequency based on the 5 MHz quartz oscillator controlled by the caesium standard, page 26, but say 60 MHz lower or higher than the nominal frequency of the klystron, and the difference, selected by the harmonic mixer, is applied to the phase-sensitive detector whose reference frequency of exactly 60 MHz is derived from the 5 MHz oscillator. When the two inputs to the phase-sensitive detector are equal in frequency but in quadrature, the output is zero, and it becomes positive or negative as the phase of the signal from the harmonic mixer changes one way or the other. This error signal, added to the permanent d.c. reflector voltage, restores equality of frequencies.

It seems that $2\,e/h$ can be measured by the method described above with a probable uncertainty less than 1 in 10^6.

Further reading:

Parker, W. H., Langenberg, D. N. and Denenstein, A. 1967. ' On the use of the AC Josephson effect to maintain standards of electromotive force '. *Metrologia*, **3**, 89–98.

Petley, B. W. 1969. ' The Josephson effects '. *Contemp. Phys.*, **10**, 139–158.

Petley, B. W. and Morris, K. 1969. ' Simple apparatus for the observation of the a.c. Josephson effect '. *J. sci. Instrum.*, **2**, 649–651.

Petley, B. W. and Morris, K. 1970. ' Measurement of $2\,e/h$ by the a.c. Josephson effect '. *Metrologia*, **6**, 46–51.

5. *The gyromagnetic ratio of the proton; the ampere*

Physicists want to know the gyromagnetic ratio of the proton as accurately as possible because it influences evaluation of a number of other important constants of physics. To electrical engineers, and others interested in realizing and preserving electric units, it is an important quantity, because the apparatus needed to determine it is also suitable for maintenance of the ampere.

We call gyromagnetic ratio of a particle the quotient m/L of its magnetic moment m and its angular momentum L, and we denote it by γ. According to the principles of quantum mechanics, the axis of spin of a particle in a magnetic field of flux density B is aligned either parallel or anti-parallel to the direction of B. If the magnetic moment is m, the magnetic energies in the two positions are $-mB$ and $+mB$, and a transition from one state to the other corresponds to a change of energy $2mB$. Each transition is accompanied by emission or absorption of radiation of energy equal to the difference of the energies of the two states. If the frequency and angular frequency of the radiation are denoted by f and ω, and Planck's constant by h, the energy absorbed or emitted is also equal to hf or $h\omega/2\pi$. Equating this expression to $2mB$, and remembering that the angular momentum L of the particle is $h/4\pi$, we find

$$\omega = \gamma B. \tag{5.2}$$

The classical picture, easier to visualize and equally instructive, is that of a particle possessing mass and charge, and spinning about an axis. The mass and spin endow it with angular momentum and the charge and spin with magnetic moment. The axis of the spinning particle, inclined at some angle θ with the direction of the magnetic field B, experiences a torque $mB \sin \theta$, but the angular momentum prevents it from at once aligning itself with the field, and causes it instead to precess about the direction of the field as a spinning top does about the vertical, fig. 5.6. If we equate the torque $m \times B$ to the rate of change of angular momentum $\omega L \sin \theta$, fig. 5.7, we find the same value as before for the angular frequency of this 'Larmor precession'. The ratio γ of m to L would be equal to $e/2M$ if to each element of mass there was a proportional element of charge, but as the measured value of γ is about 5·58 times that predicted, we are forced to abandon our over simple picture of the particle, and regard γ as an experimental constant to be determined for each nucleus.

If the protons are provided by matter in bulk, if for example they are nuclei of hydrogen atoms in the molecules of a volume of water, collisions are caused by the thermal energy, which is $kT/2$ per degree of freedom, and far exceeds the magnetic energy mB even in strong fields. The collisions produce a decrease, on the average, of the angles of precession, and force the axes of spin to set themselves parallel to the field. The

58

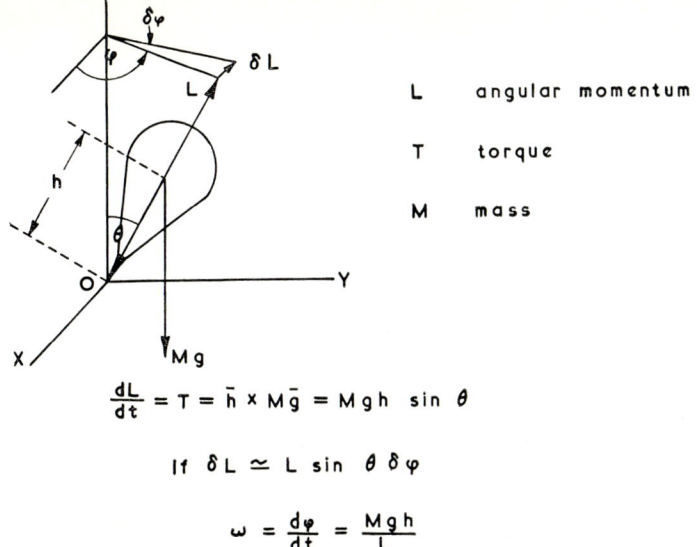

L	angular momentum
T	torque
M	mass

$$\frac{dL}{dt} = T = \bar{h} \times \bar{Mg} = Mgh \ \sin \theta$$

$$\text{If } \delta L \simeq L \sin \theta \ \delta \varphi$$

$$\omega = \frac{d\varphi}{dt} = \frac{Mgh}{L}$$

Fig. 5.6.   Precession of a spinning top in a gravitational field.

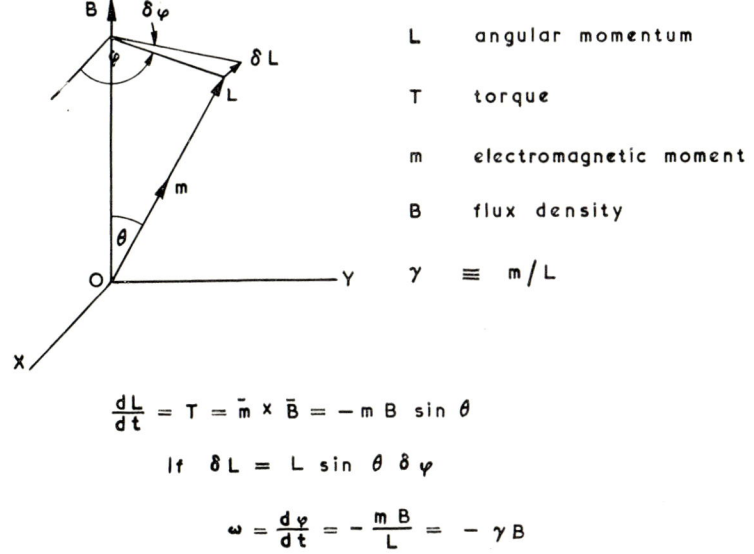

L	angular momentum
T	torque
m	electromagnetic moment
B	flux density
γ	$\equiv m/L$

$$\frac{dL}{dt} = T = \bar{m} \times \bar{B} = -mB \ \sin \theta$$

$$\text{If } \delta L = L \sin \theta \ \delta \varphi$$

$$\omega = \frac{d\varphi}{dt} = -\frac{mB}{L} = -\gamma B$$

Fig. 5.7.   Precession of a spinning charged particle in a magnetic field.

process is subject to a ' relaxation time ' which differs from liquid to liquid, and is about 3s for water.

Formula (5.2) suggests a simple method of measuring the gyromagnetic ratio: we produce a calculable magnetic field by sending a known current in a coil of measurable dimensions, somehow we arrange for precession to occur, and we observe the frequency of the e.m.f. which the precessing magnetic moments of the particles induce in a coil surrounding the vessel containing the liquid.

In fig. 5.8 the magnetic field is produced by the long coil and by an auxiliary coil of a few turns which makes up for the finite length, about 1 m, of the long coil, and ensures uniformity of field throughout the volume of the water container, a spherical glass or quartz shell 40 mm in diameter, placed at the centre of the apparatus. Round the sphere is a detector coil tuned to the frequency appropriate to the magnetic field,

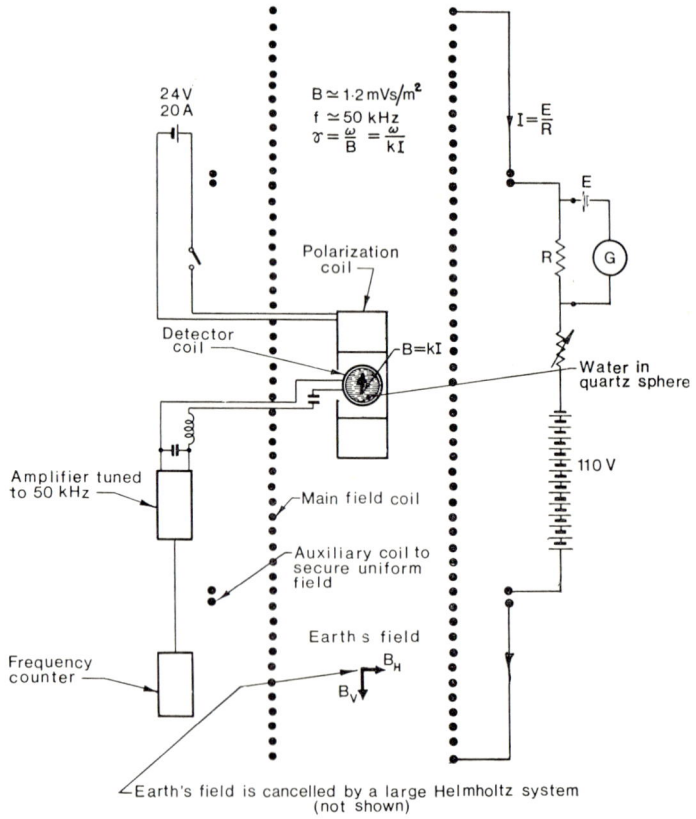

Fig. 5.8. Circuit for measurement of the gyromagnetic ratio of the proton by the method of free precession in a weak magnetic field.

60

about 50 kHz; the tuned circuit is followed by a narrow-band amplifier and a counter to measure the duration of 10, 20 or 40 thousand periods of the signal.

The current in the field-forming coils flows also in a resistor R of nominal value 1 Ω and is manually or automatically adjusted for the p.d. at the potential terminals, page 50, to balance the e.m.f. of the Weston cell E. The current is thus nominally E/R, but as explained in chapter 3, page 34, E/R can be determined in amperes by the current balance.

A few seconds after the standard field has been established, the axes of the protons will have acquired, on the average, a small bias in its direction, and there is no signal. The bias is small because the thermal energy, which tends to produce disorder, so greatly exceeds the magnetic energy, which tends to cause the alignment. The largest signal is obtained if to start with the polarization of the protons is large, and if the axes of spin are biased at right angles to the standard field. This alignment is effected by a large current maintained for 4 or 5 seconds in the polarizing coil, to produce a field about 100 times as large as the standard field and a correspondingly greater polarization of the protons. If after that time the polarizing current is switched off quickly, precession about the standard field, which alone remains, starts at almost the optimum angle of $90°$ and, because of the increased polarization, produces an initially large signal which decays in 2 or 3 seconds, giving ample time for accurate measurement of its period.

The standard field B is calculated from the current and the dimensions of the two field-forming coils which, like the coils of the current balance or of the standard inductor, are wound with a single layer in helical grooves cut on the surfaces of cylindrical formers. The Earth's field must be removed or allowed for. A simple way of eliminating it is to send the requisite current in two pairs of Helmholtz coils, one with the axis vertical, the other with the axis along the horizontal component of the Earth's field. The currents in each pair can be adjusted by means of a small magnetometer located, for that adjustment only, at the centre of the apparatus, in place of the spherical container. Cancellation need not be perfect, for if there are residual components B_1 along and B_2 at right angles to the standard field, B_1 is eliminated by reversing B and taking the mean of the two frequencies, whereas B_2 forms with B a resultant field in a direction slightly inclined to that of B. This resultant field differs in magnitude from B as the hypotenuse differs from the long side of a right-angled triangle of small third side, i.e. by a relative amount $B_2{}^2/2B^2$. This ratio is less than 1 in 10^7 when B corresponds to about 50 kHz, even if the cancellation by the Helmholtz coils is good only to 1% of the Earth's field. There is no difficulty in cancelling to $0\cdot1\%$.

The value of γ obtained in this way is not the gyromagnetic ratio proper, which refers to an isolated proton, because the electron of the hydrogen atom screens the proton from the field B. The difference is about 26 in 10^6 and has been calculated and also confirmed by experiment. For maintenance of the ampere it is neither necessary nor even desirable to apply this correction, nor is it necessary to apply a correction for the paramagnetism of the liquid and the shape of the container—it is simpler to agree to use water in a spherical vessel so that the results of different laboratories may be strictly comparable.

A measurement of γ in terms of E/R by this method is subject to a probable uncertainty of perhaps 1 in 10^6, mostly due to systematic errors which do not come in when the apparatus is used in reverse to check the constancy of E/R. For this purpose it is not even necessary to have previously determined γ to the above accuracy, it is sufficient to agree on a value, or just to state the particular figure adopted for any one measurement.

The first accurate measurement of γ by a method generally similar to that described above was made by the National Bureau of Standards, USA, in 1957.

The value appears to be close to 2·675 127 rad/s T, the units being derived from (5.2); an equivalent form sometimes useful is 42·576 MHz/T, the units in this case being derived from the actual method of measurement. Whichever form is used, it is important to remember that γ depends on the unit of current, not the unit of potential.

Determination of γ in terms of the ampere rather than in terms of the units of e.m.f. and resistance of the laboratory, as maintained by Weston cells and standard resistors, is less accurate, for the ampere itself, page 34, is subject to a probable uncertainty of some 4 in 10^6. It may be asked why that should be so, since in each case the vital components of the apparatus are coils of wire wound on cylindrical formers. The answer is that calculation of magnetic flux density at the centre of a long coil is inherently more accurate than that of force between two coils, for the latter is very dependent on the diameters as well as the pitch of the windings, whereas the former involves mainly pitch, with diameters appearing only in so far as the coil is not infinitely long. The outcome is that the ampere can be checked more accurately than it can be realized. From what was said above of the accuracy of control of the volt by the Josephson effect, it is clear that the state of affairs for the volt is similar, however it may be realized.

The 'weak field' method described above is not the only one available for determining the gyromagnetic ratio of the proton. Another, the 'strong field' method, will be outlined because the two in conjunction yield not only the gyromagnetic ratio but also the ampere itself, thus rendering superfluous dynamometers, like the current balance, with which the unit has hitherto been realized.

The magnetic field of about 1 T is that of a large magnet of the type used for nuclear magnetic resonance. The water container, much smaller, say 10 instead of 40 mm in diameter, is in the centre of the gap, as in fig. 5.9. Rather than observe free precession it is more convenient

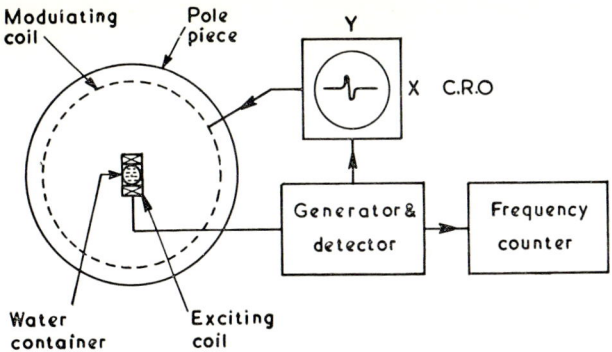

Fig. 5.9. Measurement of the gyromagnetic ratio of the proton in a strong magnetic field.

in this case to stimulate transitions by a continuous-wave radio-frequency field, of approximately 42·5 MHz if the flux density is 1 T, applied by a coil surrounding the water container and with its axis parallel to the pole pieces. When the applied radio frequency is equal to the precessional frequency the transitions are a maximum, in other words resonance occurs, which can be detected either by the reaction on the exciting circuit, or else by observing the signal induced by the precessing protons in a second coil also surrounding the container, but with its axis perpendicular to that of the exciting coil though still parallel to the pole faces. Measurement of the frequency of the alternating field offers no problem since it is applied continuously, but it is more convenient to keep that frequency fixed at approximately the correct value, and produce resonance by modulating the magnetic field of the magnet to a very small extent at 50 Hz, by means of two coils inside the gap against the pole faces, and to display on the screen of an oscilloscope the induced or the absorption signal vertically, and the 50 Hz current horizontally. The small correction to B is estimated from the position of the resonance peak along the time base. Alternatively, modulation can be effected by a unidirectional current increasing linearly with time, provided by the ordinary time base of oscilloscopes.

The magnitude of the magnetic flux density B, required for the calculation of γ from (5.2), is obtained from the force exerted by the field on a known current. For the purpose a coil of rectangular shape, long enough for one short side to be well clear of the pole pieces when the opposite side passes through the region of the gap previously occupied by the water container, is suspended from one end of a balance beam whose other end carries a counterpoise, and the change of force due to reversal of the current in the coil is measured by means of a weighing mass removed from or added to a scale pan hanging from the same suspension as the coil. Just as for the weak field method or the current balance, the current is controlled by means of a Weston cell of e.m.f. E and a resistor

63

of resistance R, thus is equal to E/R in terms of the laboratory units. The change of force on reversal of the current is $2a(B-B')E/R$ where B' is the small stray flux density at the upper end of the coil, fig. 5.10. This stray field, as well as the field in the gap, is mapped by relative precessional frequency measurements or otherwise, averages are taken, and the final result is referred to the flux density at the centre of the lower side of the rectangle.

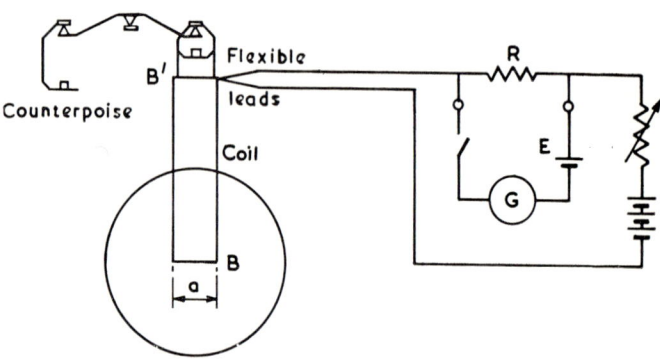

Fig. 5.10. Determination of flux density at centre of gap of magnet used to measure precessional frequency of protons.

Measurements of the gyromagnetic ratio of the proton by the strong field method have hitherto suffered from an uncertainty at least as high as that of measurement of current by current balances. It is interesting to note, however, that in the weak field method we find

$$\omega_w = \gamma B_w$$
$$= \gamma \mu_0 H$$
$$= \gamma \mu_0 k I$$
$$= \gamma \mu_0 k E/R\alpha \qquad (5.3)$$

where α is the ratio, close to unity, of the ampere and the laboratory unit maintained by means of resistors and Weston cells, whereas in the strong field method

$$\omega_s = \gamma B_s$$
$$= \gamma k'/(E/R\alpha) \qquad (5.4)$$

where k depends on the dimensions of the field-forming coils, and k' on the rectangular coil, the weighing mass, the acceleration due to gravity, and the correction for B'. From (5.3) and (5.4) we obtain γ by multiplication and α by division, thus realizing the ampere as well as determining the gyromagnetic ratio.

64

If the strong-field apparatus eventually proves at least as accurate as the current balance, it should be preferred because the magnet can also be employed to measure another fundamental constant, viz. the ratio e/m of nuclei, and for general work on nuclear magnetic resonance.

Further reading:

Bender, P. L. and Driscoll, R. L.   1958.   ' A free precession determination of the proton gyromagnetic ratio '.   *IRE Trans.* I—**7**, 176–180.

Capptuller, H. A.   1963.   ' The proton gyromagnetic ratio as a nuclear standard '.   *Nuclidic Masses* 1963, 105–112.

Vigoureux, P.   1962.   ' A determination of the gyromagnetic ratio of the proton '.   *Proc. Roy. Soc. A*, **270**, 72–89.

6. *The calculable capacitor; the ohm*

Whereas the volt and the ampere can be maintained by reference to atomic constants more accurately than they can be realized by reference to the classical laws of electromagnetism, and whereas until some years ago the same was true of the ohm, the latter has been transferred to a class of its own by the advent of the calculable capacitor, which seems as convenient for watching the annual drift of resistors as it is for realizing the unit.

The present position then is that three precise methods are available to check the material standards, whereas on account of Ohm's law only two are sufficient.   This redundancy has of course the advantage of providing a test of the precision claimed for the three methods.

7. *The speed of light; the metre and the second*

We have been concerned hitherto with the speed of light only in so far as it affects realization of the farad, and indirectly of the ohm, by the calculable capacitor, and have seen that the uncertainty of its value formed the largest contribution to the errors associated with the realization of the ohm.   There is however a much more fundamental reason for wanting to know it to 1 in 10^8 rather than to the present 3 in 10^7.   If we accept that it is a universal constant, we should in principle be entitled to define the unit of length in terms of it and of frequency, rather than in terms of wavelength.   Thus in the same way that frequency is now used to maintain the volt, it could serve to realize and maintain the metre itself, and therefore also the ampere through the gyromagnetic ratio of the proton.   The gain is of course that frequency is measurable to 1 in 10^{12} whereas the metre as at present defined is realizable to perhaps 1 in 10^8.

This attractive possibility is however subject to the condition of preserving continuity in the size of the units, a condition which in turn requires that the speed of light be known to 1 in 10^8 before a value is adopted and fixed for good.   It is thus not surprising that work aimed at securing that accuracy should be in progress in a number of places.   As however not only the details, but the methods even, are by no means finalized, it is sufficient here to confine ourselves to the principles underlying the investigations.

The second is adequately fixed by the frequency of transitions of caesium atoms, and the metre is defined in terms of the wavelength of the orange line of krypton 86, approximately 606 nm. It is not at present possible to measure the frequency of the krypton line directly in terms of the caesium frequency, but a method proposed in the United States of America consists in exciting an optical laser in two modes, of frequencies say f_1 and f_2 where f_2-f_1 might be in the gigahertz range. In addition to the outputs at λ_1, λ_2 corresponding to f_1, f_2 the laser also supplies energy at the beat frequency f_2-f_1 which can be compared with the caesium frequency, whilst λ_1 and λ_2 can each be determined in terms of the krypton wavelength. Since

$$f_2-f_1=c\left(\frac{1}{\lambda_2}-\frac{1}{\lambda_1}\right) \tag{5.5}$$

the speed of light follows.

The NPL is considering an intriguing alternative: if it were possible to lock to the caesium frequency a laser wave of length comparable to that of the krypton wave, a comparison of the two optical wavelengths by interferometry would yield the desired accuracy. Locking an optical frequency to the 5 MHz controlled by the caesium beam, page 26, is however not easy and cannot at present be effected in a single step.

A 2 mm klystron generator can however be locked by a synthesizer fed by the 5 MHz crystal, much as was described on page 57 in connection with the Josephson effect. The klystron output is then applied to a mixer also fed from a 337 μm HCN laser, and the rectified beat between a harmonic of the first and the laser stabilizes the latter by acting on a piezoelectric plate controlling the length of the cavity. Even then the step of over 500 to 1 to reach the optical frequency is too long for comparison by interferometry, which involves corrections proportional to the square of the wavelength, negligible at 600 nm, but too large at 337 μm. One possibility under study at NPL is to introduce intermediate steps as follows: lasers working on water vapour lines at 118 and 28 μm, and a carbon dioxide laser of approximately 10 μm, are stabilized in succession, and a helium-neon laser of 632·8 nm is modulated by the CO_2 laser, thus yielding sidebands of wavelengths suitable for interferometry, spaced in frequency by exactly twice the frequency of the stabilized CO_2 laser.

If we denote the frequencies of the two lasers by f_H and f_C and the wavelengths of the sidebands by λ_S and λ_D, we have

$$f_H+f_C=c/\lambda_S$$
$$f_H-f_C=c/\lambda_D$$

whence

$$2f_C=c\left(\frac{1}{\lambda_S}-\frac{1}{\lambda_D}\right). \tag{5.6}$$

Since f_C is known in terms of the caesium frequency, and since λ_S and λ_D can both be measured in terms of the krypton wavelength, possibly through the intermediary of the helium-neon wavelength, the speed of light results from (5.6) to an accuracy depending mainly on that of the measurement of the two wavelengths.

Several of the operations involved above, especially stabilization of the intermediate lasers and modulation of the helium-neon laser, are not easy to achieve, but, if the difficulties can be overcome, the speed of light will be determined with a precision of 1 in 10^8 or better. If some value for it were then agreed upon, the wavelength definition of the metre could be replaced by one stating the frequency of the line, and the metre would be defined with the same precision with which frequency can be measured.

The change would in no way reduce the need for interferometry, which would be applied as it is now to measurement of lengths and displacements. The only difference would be that the wavelength of the light, more often than not a laser line, would in theory have been calculated from its stated frequency and the speed of light, and could be specified to more significant figures than at present, even though the interferometer employed for the particular calibration might not be capable of making full use of that precision.

Further reading:

Froome, K. D. and Essen, L. 1969. *The Velocity of Light and Radio Waves.* Academic Press, Lond. & N.Y.

CHAPTER 6
comparison of standards

1. Resistors and Weston cells

THE ohm, the volt and the ampere are maintained from day to day in standards laboratories by reference groups of resistors and of Weston cells. We have seen, on the other hand, that resistors and cells form part of equipment which serves to realize or maintain electric and magnetic units. It would not be safe to employ resistors or cells of the reference groups for this purpose, and in any case the resistors needed are often of different construction, since they are expected to carry, without over-heating, currents of the order of 1 A, or to be suitable for a.c. as well as d.c., properties not possessed by the reference standards nor indeed required by them. It is however necessary to compare these special resistors, and also the cells employed in the realization and maintenance apparatus, with the reference standards. Moreover much measuring equipment includes, as components, resistors and cells, which also need calibrating at regular intervals.

Many instruments have been employed for the purpose, of which we shall mention only a few. Some, devised in the nineteenth century, are well known and still fulfil the most stringent requirements set by the desire for ever increasing precision, whilst the others are recent develop-ments which look so promising as to suggest that they may well displace some of the earlier methods of comparing resistance standards. These instruments are, on the one hand, the Wheatstone bridge, from which the Kelvin double bridge is derived, the resistance ' build-up ' or ratio box, and the potentiometer; on the other hand, the inductive divider or transformer bridge, and the direct current comparator.

2. The Wheatstone bridge

The Wheatstone bridge is a network consisting of four resistors, P, Q, R, S connected end to end to form a closed loop, fig. 6.1, with a battery connected to two opposite junctions, say those of P, Q and of R, S, and a galvanometer between the other two. The condition for bridge balance, i.e. for no current in the galvanometer, is that QR/S shall be equal to P. Some laboratories, the NPL of the UK for example, compare resistance standards of nominal value 1 Ω by the substitution method. The bridge is assembled from four resistors of the type described on page 49 and in fig. 5.1, one of which, P, is one of the standards to be compared, and the others, Q, R and S are of similar type, but of nominal values say 10, 10 and 100 Ω. The current leads of the resistors are shaped to rest in mercury cups on copper blocks on four insulating pillars in an oil bath, each block having two cups and a terminal; thus the bridge is easily altered to suit requirements. Either Q or S is shunted by a resistance box which serves to balance the bridge, and from whose reading the difference between two resistors P_1 and P_2 of nominal value 1 Ω can be obtained. It is not necessary to know the exact values of Q, R and S, provided they are close to nominal.

If, as is general practice, the 1-ohm standards are provided with potential terminals, it is possible, by repeating the measurements with the battery and galvanometer connected to the potential terminals of P instead of to the current terminals on the blocks, fig. 6.1, to deduce from the two sets of values of the shunt resistors the resistance of P between its potential points.

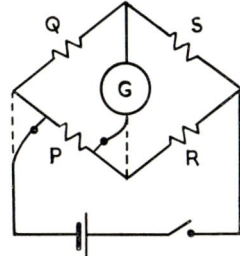

Fig. 6.1. Wheatstone bridge for comparison of standard resistors.

By suitable choice of the ratio arms, resistors of 10 Ω, 100 Ω, etc., can also be compared on the Wheatstone bridge.

3. The ' build-up '

The substitution method serves to compare resistors of the same nominal value, but a standardizing laboratory must also know its 10 ohm, 100-ohm, etc., resistors in terms of its 1-ohm standard. For this purpose a resistance box is made of eleven 10-ohm resistors in series. The intermediate nine can be connected in series-parallel by two links to yield a resistance of 10 Ω, which is compared with the first and the last 10-ohm resistors. From this simple comparison the resistance of the first ten resistors in series is obtained in terms of the resistance of the eleventh resistor to a part in 10^7. A different method has been used by Hamon to obtain 1 ohm by connecting ten 10-ohm resistors in parallel, and 100 ohms by connecting them in series. Hamon has improved this build-up by inserting compensating resistors in the potential leads, thus making it possible to measure 100-ohm resistors in terms of a 1-ohm standard to an accuracy better than 1 in 10^7.

4. *The Kelvin double bridge*

Resistors of 0·1, 0·01 and 0·001 ohm may be compared on the Kelvin bridge, fig. 6.2, in which the galvanometer, instead of being connected to the junction of the current leads of *P* and *R*, as in the Wheatstone bridge, is connected to the junction of auxiliary resistors α and β whose other ends go to the potential leads of *P* and *R*. The current leads of *P* and *R* can be joined by a stout removable link *d* of low resistance. Various methods of balance may be used, in one of which successive balances are

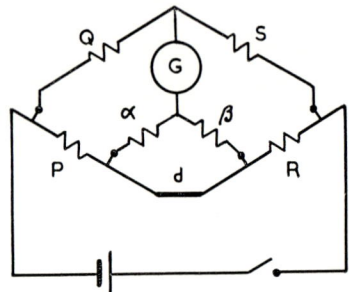

Fig. 6.2. Kelvin double bridge for comparison of resistors.

obtained with the link in and the link out by adjusting shunts across *Q* and α alternately. At balance

$$\frac{P}{R} = \frac{Q}{S} = \frac{\alpha}{\beta} \qquad (6.1)$$

but it is not necessary to know α and β, which include one potential lead of *P* and one of *R* respectively. With these two balances, the other two potential leads are not eliminated, but they can be measured or allowed for by a third balance. This bridge is also often used to compare 4-terminal 1-ohm resistors, as it avoids the change of connections necessary with the Wheatstone bridge to obtain the resistance between the potential terminals.

5. *The potentiometer*

The potentiometer, fig. 6.3, is likewise an instrument which, although devised years ago, is nevertheless still convenient for comparing resistors and still almost always used, in one form or another, to compare Weston cells. Its main component is a resistor to which contact can be made, ideally at any desired point, but more often at quite small intervals. This resistor carries a current which can be adjusted and subsequently kept constant. A Weston cell, galvanometer and key are connected between one end of the resistor and the tapping point, which is set to a value of resistance corresponding to the numerical value of the e.m.f. *E* of the cell, and the current in the resistor is adjusted for zero current in

70

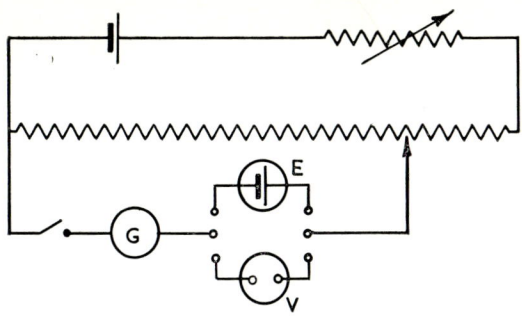

Fig. 6.3. Principle of potentiometer.

the galvanometer. The readings of resistance corresponding to any other position of the tapping point then give difference of potential in volts, thus any source of potential difference V can be measured by switching the tap and galvanometer to it, and adjusting the position of the tap for balance.

To compare a number of Weston cells, they are in turn put in opposition against one cell, whose value need not be known but should not change significantly during the measurements, and the differences of e.m.f. are observed. They are in general small, since Weston cells do not differ much from one another in e.m.f., but there are potentiometers designed especially for accurate measurement of small e.m.f.s.

The potentiometer is also used for comparing resistors. If current is maintained in resistors connected in series, the potential differences between the potential terminals of each can be read on the potentiometer and will be proportional to the resistances. This method automatically gives the resistance between the potential points of 4-terminal resistors.

6. *Inductive potential dividers*

Although normal national and international practice has been to compare standard resistors with direct current, the ohm is now almost always realized by an alternating current method from a known inductor or capacitor. It is only necessary to use for the purpose resistors of low reactance and to verify, by measurements at a few frequencies, that at the low frequency employed for the realization, the resistance has the same value as with d.c., or at worst differs from the d.c. value by only a small, known amount. There would be no objection therefore to the use of a.c. for comparing 4-terminal resistance standards provided they also were designed to fulfil the above condition. Recent work has shown that with such resistors the a.c. inductive potential divider provides an accuracy comparable to, or even better than, that obtainable with the d.c. Wheatstone bridge.

In principle the inductive divider, whose accuracy owes a great deal to recent development at the NPL, is an autotransformer with a core of low loss, high permeability material and a winding in ten sections, whose flux linkages are equal to one another to a few parts in 10^8, or even in 10^9. In order to secure a large range of voltage ratio, several transformers, each with a winding made from 10-strand cable, are used, to give nine decades with tappings brought out to studs of dials, as in resistance boxes. In principle, each successive decade is fed from the previous one, but there are refinements which reduce errors further still.

We saw above that the Kelvin double bridge is convenient for comparison of 4-terminal resistors with d.c.: it is also suitable for use on a.c. with inductive dividers, which take the place of the ratio arms Q, S and α, β of fig. 6.2 and secure a considerable simplification in arithmetic, since the dials of the main divider in fig. 6.4 give the ratios directly, whereas

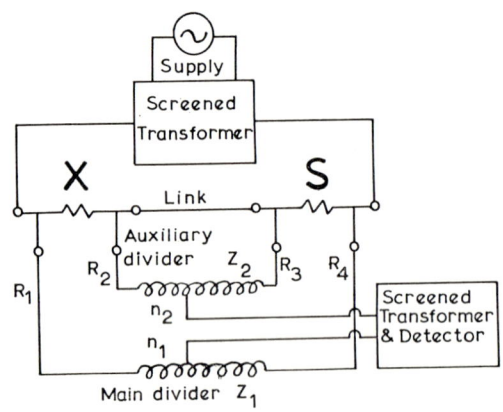

Fig. 6.4. Comparison of resistors by Kelvin double bridge, with inductive dividers in place of the resistive main and auxiliary ratio arms.

with the d.c. bridge the effect of the shunt on the arm Q has to be computed. Moreover the high input impedance of inductive dividers renders superfluous the procedure used with the d.c. bridge to eliminate the effect of leads. Once the bridge has been balanced with link in and link out, we have by analogy with (6.1)

$$\frac{X}{S} = \frac{n_1 Z_1 + R_1}{(1-n_1)Z_1 + R_4} = \frac{n_2 Z_2 + R_2}{(1-n_2)Z_2 + R_3}. \tag{6.2}$$

In practice R_1, R_2, R_3, R_4 can be made sufficiently small compared with Z_1, Z_2 for them to be neglected in (6.2) and we have

$$X = S \frac{n_1}{1-n_1} \tag{6.3}$$

with an uncertainty not exceeding 1 in 10^7.

72

It is even possible, by ganging the dividers for equal ratios, to secure balance without removing the link, for, from (6.2), n_1 is equal to n_2 if the resistances of all the leads can be neglected.

As in all a.c. bridges the quadrature components due to phase defects, if any, of the resistors need to be compensated for perfect balance. A variable shunt capacitor across X or S, fig. 6.4, will provide the necessary adjustment, which can, alternatively, be effected by means of a small variable mutual inductor whose primary is in series with X and whose secondary is in one of the potential leads of X.

7. Direct-current comparator

The direct-current comparator, also based on accurately wound decade transformers, is more promising still, and in a way more satisfactory for present requirements, since it takes us back to measurement with direct current. It was developed by N. L. Kusters and his colleagues of the National Research Council of Canada, and is produced by a firm in that country. The principle is that when the ampere-turns of two windings on the same magnetic core are equal and opposite, the flux in the core is zero. In fig. 6.5 are shown two independent d.c. supplies, one for the

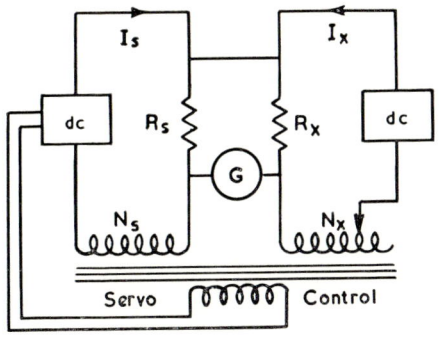

Fig. 6.5.   Direct current comparator applied to comparison of resistors.

standard, S, and one for the unknown, X, branches of the bridge. If a current I_s is maintained in the S branch, and if a current I_x in the X branch be adjusted for the galvanometer to read zero, $R_x I_x$ is equal to $R_s I_s$. If moreover the turns N_x be adjusted for magnetic balance, $N_x I_x$ is equal to $N_s I_s$. These two relationships are equivalent to R_x being equal to $R_s N_x / N_s$. Features of the bridge are firstly the servo system which ensures magnetic balance by automatic adjustment of I_s, and secondly the facility provided for N_x to yield the resistance R_x directly in ohms although the value of the standard R_s may, as is normally the case, differ from 1 ohm by a small, known value.

Flux in the core, in other words a difference between $N_s I_s$ and $N_x I_x$, is detected by means of two auxiliary windings on the core. The modulator winding, fed with a.c., varies the permeability of the core over the cycle, and as a result an e.m.f. at double the applied frequency is induced in the detector winding unless there is no unidirectional flux in the core. This harmonic output is made to act on the power supply of the S branch to alter I_s until magnetic balance is secured. Thus as N_x is adjusted manually to bring the galvanometer current to zero, I_s is automatically adjusted for magnetic balance by the feedback current, and R_x is given by the relation above.

In the bridge as produced, the windings N_s consist of a fixed 1000-turn coil, and of decades which provide the equivalent of 11·1110 turns in either polarity, so that any departure of the standard 1 Ω resistor R_s from nominal may be compensated. When the seven decades of N_x have been adjusted for balance, the numerical value of R_x in ohms is therefore simply a decimal submultiple of the reading of N_x.

Thus the direct-current comparator yields the resistance between potential points directly in ohms with a single manual operation, whereas the d.c. Wheatstone or Kelvin resistance bridges require at least two balances and a computation of the effects of shunts. There is also a difference in the sharing of power between resistors under comparison. When resistors of say 1 Ω and 10 Ω are compared on the bridges, they take equal currents and the high resistance dissipates ten times as much power as the lower. In the comparator the currents are shared in the inverse ratio of resistance, and therefore the greatest power is dissipated in the low resistance. Presumably the ideal would be equal powers, but in some cases the sharing in inverse proportion may be the better compromise.

The direct current comparator is a very recent development, and will need to be compared with the bridges described above to verify that the great advantages it appears to have are not marred by defects hitherto unnoticed, but there is every likelihood that it will revolutionize comparison of resistors, for it is as well adapted to comparison of small resistances with a 1 Ω standard as to inter-comparison of 1 Ω resistors.

8. d.c./a.c. transfer

Electrical energy is mostly used in the form of alternating current, thus although the realization of the units is effected with direct current, or with alternating current at low frequency, and although reference standards have in the past been compared by direct-current methods, working standards for alternating currents, especially for power frequencies, are a necessity.

Resistors do not present much difficulty because it is possible to design them so as to keep the effects of reactance and eddy currents negligible, but since the Weston cell is an essentially d.c. device, some instrument is required which can be compared with a cell, and which reads the same with a.c. and d.c.

74

Methods vary with laboratories. The NPL uses an electrostatic voltmeter of very special construction which, in conjunction with resistive or capacitive potential dividers, effects a transfer from the Weston cell to the higher voltages employed for use and transmission of electricity.

More recently the NPL and other laboratories have developed instruments of the thermojunction type, now known as thermal convertors, which can serve for d.c. to a.c. transfer not only at 50 or 60 Hz but at much higher frequencies. These instruments have already started to play their part in the calibration of a.c. working standards and there seems every likelihood of their being used more and more.

Power being for economic reasons an important quantity in the electrical industry, it is natural that laboratories should be interested in the calibration of instruments by which it can be measured in terms of the volt and the ampere. These instruments take the form of wattmeters of the electrodynamic or, less frequently, of the thermal type. To calibrate them the NPL uses an electrostatic wattmeter in the form of a quadrant electrometer standardized in terms of the ohm and volt. The former is provided by standard resistors, and the alternating volt is measured by the electrostatic voltmeter.

9. *International comparisons*

As the resistors and cells of the several national laboratories do not necessarily change by equal amounts in the course of years, national units are liable to drift apart, especially among those countries which do not yet have facilities for correcting the drift of their material standards by means of absolute measurements or by reference to atomic constants. Since on the other hand it is desirable that the differences between national units be known, the International Committee of Weights and Measures has arranged for periodical comparisons of the national units of resistance and tension. These comparisons are made every three years or so by the International Bureau of Weights and Measures at Sèvres.

Each national laboratory, on some agreed date, sends a few resistors and cells to the International Bureau, which compares them all with its own material standards. On completion of its measurements the Bureau returns the standards to their owners, who measure them again and inform the Bureau, which then calculates the differences between the national units and its own, and publishes the results.

It is not usual for the participating countries, of which there were ten in 1968, to adjust their units as a consequence, except at rare intervals and by international agreement, because continual small adjustments would upset the continuity of calibrations in each single country. Indeed since 1948, when adjustments of a few parts in 10^4 were made, there had been no general change until January 1st, 1969. On that day all participants adjusted their units according to the then best available absolute determinations. The adjustments made by the United Kingdom were an increase of the ohm by 3·7 in 10^6 and a decrease of the volt by 13 in 10^6.

International comparisons of material electric standards, although not meant primarily for the benefit of pure science, but rather for that of the electrical industry generally, have nevertheless proved of great value to physicists interested in improving the accuracy of our knowledge of fundamental constants. Many of these constants, e.g. the gyromagnetic ratio of the proton or the ratio $2e/h$, are conveniently expressed in the first instance in terms of the ohm and the volt or both, as maintained by resistors and Weston cells, for expression in terms of the true ohm or volt would entail ' absolute ' determinations of these units, for which the necessary equipment might not be available where the fundamental constant is being measured. Measurements made by the International Bureau facilitate the comparison of values of the same constant determined in two or more laboratories separated by continents or oceans, and relieves them from the onerous task of organizing their own independent comparisons of material standards. The saving of time and effort, and the increased accuracy thus afforded, illustrate the value of international collaboration in science.

Further reading:

Deacon, T. A. and Hill, J. J. 1968. ' Two-stage inductive voltage dividers '. *Proc. Instn elect. Engrs*, **115,** 888.

Hamon, B. V. 1954. ' A 1–100 ohm build-up resistor for the calibration of standard resistors '. *J. sci. Instrum.*, **31,** 450.

Hill, J. J. 1965. ' Calibration of d.c. resistance standards and voltage ratio boxes by an a.c. method '. *Proc. Instn elect. Engrs*, **112,** 211–217.

MacMartin, M. P. and Kusters, N. L. 1966. ' A direct current comparator ratio bridge for four-terminal resistance measurements '. *Trans. Inst. elect. electron. Engrs*, 1M—**15,** 212.

Wilkins, F. J., Deacon, T. A. and Becker, R. S. 1965. ' Multijunction thermal convertor: an accurate d.c./a.c. transfer instrument '. *Proc. Instn elect. Engrs*, **112** (4), 794.

FUNDAMENTAL CONSTANTS

The latest assessment of the 'best' values of atomic constants was made by Taylor, Parker and Langenberg. We give in Table 7.1 below the values, taken from their Table 32, of the few constants mentioned in this book, together with the assessors' estimate of the standard uncertainty (68·3% probability) and the units in which the constants are expressed. To obtain probable uncertainties (50%) used throughout this book, it is sufficient to divide the figures of column 4 by 1·5. To express N and F in terms of the base unit, the mole, reduce by 3 the exponents in the last column.

Values of other constants, consistent with the above, are given in Table 32 of *The Fundamental Constants and Quantum Electrodynamics*, Taylor, B. N., Parker, W. H. and Langenberg, D. N. Academic Press: London 1969.

Quantity	Symbol	Value	Standard uncertainty, parts in 10^6	Unit
Speed of light	c	2·997 925 0	0·33	10^8 m s^{-1}
Electronic charge	e	1·602 191 7	4·4	10^{-19} C
Planck's constant	h	6·626 196	7·6	10^{-34} J s
Avogadro's number	N	6·022 169	6·6	10^{26} kmol^{-1}
Magnetic flux quantum	$h/2e$	2·067 853 8	3·3	10^{-15} Wb
Faraday's constant	F	9·648 670	5·5	10^7 C kmol^{-1}
Gyromagnetic ratio of protons in water	$\gamma_{p'}/2\pi$	4·257 597	3·1	10^7 Hz T^{-1}
Boltzmann's constant	k	1·380 622	43	10^{-23} J K^{-1}

Table 7.1. 'Best' values of some atomic constants.

THE WYKEHAM SCIENCE SERIES

THE WYKEHAM TECHNOLOGICAL SERIES

All orders and requests for inspection copies should be sent to the appropriate agents. A list of agents and their territories is given on the verso of the title page of this book.